FAO FISHING MANUALS
ECHO SOUNDING AND SONAR FOR FISHING

FAO Fishing Manuals

Echo Sounding and Sonar for Fishing

Published by arrangement with the
Food and Agriculture Organization of the United Nations
by **Fishing News Books Ltd**
1 Long Garden Walk, Farnham
Surrey, England

British Library C I P Data

Echo sounding and sonar for fishing.—(Food and Agriculture Organization. Fishing manuals).
1. Echo sounding in fishing
2. Sonar in fishing
I. Series
639'.2'028 SH344.23.E3

ISBN 0 85238 110 7

Typeset by
Information Services Division,
Brown Knight & Truscott Ltd.,
Tonbridge and London

Printed in Great Britain by
Whitstable Litho

CONTENTS

LIST OF FIGURES

INTRODUCTION

Since the introduction into fishing of acoustic methods of fish detection in the early 1950's, much effort has been devoted by scientists and industry to improving the performance of acoustic equipment and in better adapting it, and the associated techniques, to the varied requirements of different fisheries. This process has been well recorded in booklets, periodicals and proceedings of meetings and conferences, and a selection of such publications (in English), will be found at the end of this book under *Further Reading.*

It is generally accepted that specific skill is needed in order to take full advantage of a new and continuously progressing technique such as underwater acoustics for fishing. Obviously such skill can be best and most quickly acquired by systematic studies and training in conjunction with practical experience. This manual is meant to contribute to the first two steps of this learning and upgrading process by providing essential background information on physical fundamentals and equipment technology, and on their employment under different fishing conditions. The manual is intended primarily for fishermen, instructors, extension workers and fishermen trainees, but it should also be found useful by boat owners, fishery managers and administrators for selecting and installing the right equipment for a particular purpose.

By definition this manual is not a textbook on underwater acoustics for physicists or science students. Some such references are shown in the list of Further Reading and more can be found listed within those publications.

This manual is predominantly aimed at assisting developing fisheries to accelerate their technical progress. Consequently the text and layout is practice-orientated and reduces theory to the essential minimum required for understanding the principles. Emphasis is given to the technology of equipment and to its operation under various fishing conditions; the need for updating in these areas is the main reason for preparing such a manual, although similar ones are already in existence.

The manual is based on manuscripts prepared under contract with FAO by two well-known experts, Prof M Nishimura of Japan and G Vestnes of Norway. From his vast practical experience in commercial and exploratory trawling Capt M Guidicelli, FAO Fishing Technologist and Masterfisherman, has contributed the material for sonar-guided bottom trawling (*11.6*). P J Hearn, United Kingdom, edited the manuscript into one unit adding various sections, and the final form was prepared by the Fisheries Technology Service, Department of Fisheries, FAO.

Obviously, this manual could not have been produced without the active co-operation of individual experts and of the echo sounder and sonar manufacturers who provided technical information and illustrations. It must be stressed, however, that the selection of these contributors and their products, as used for illustrations, was the result of existing contacts and other 'convenience factors' and definitely should not be taken as an indication of preference for any one manufacturer over another.

As can be seen from the team of authors and contributors, the attempt was to enlist expertise from a wide geographical and technical range to cover the subject adequately and objectively. However, it is clearly not possible to be fully comprehensive within the framework of such a small booklet, and there may well be shortcomings and omissions which some readers may consider to affect its usefulness, in which event their constructive criticism is invited.

There is no doubt that echo sounding and sonar are essential for rational fishery development at all levels. Recent developments in electronics have considerably improved the performance and reliability of the equipment and have lowered costs. At the same time suppliers have given greater emphasis to inexpensive units suitable for the requirements of the so far neglected majority of small scale, poorer fishermen throughout the world. It is hoped that this manual will help these fishermen to use this equipment to the best advantage.

CHAPTER 1

SOUND IN WATER

Hydroacoustics is the general term for the propagation of sound in water. In fisheries, the term 'echo sounding' (or vertical sounding) is usually restricted to sound transmitted from a vessel and returned to it along a line straight down to the sea bed (or bottom). The equipment employing this technique is called an 'echo sounder'. Sonar is equipment based on the same principle but capable of transmitting and receiving in more or less horizontal directions. A 'netsonde' is equipment which collects information with sensors at the trawl (mainly acoustic transducers, but also others, *eg* for temperature, speed, amount of catch) and transmits it for display in the wheelhouse during trawling. Here, only the acoustic component will be discussed.

In order to understand the acoustic function of an echo sounder, sonar or netsonde, it is necessary to know something about sound propagation, *ie* the formation and spreading as well as the absorption, reflection and refraction of sound waves in water.

When we strike a gong we hear a sound. This is because the vibrations of the gong are transmitted to the neighbouring particles and these vibrations travel through the air. *Fig 1* shows high and low pressure alternating in sound waves.

If the quality of the gong is good, it will give only one single clean tone. The pressure curve will be regular and the distance between any of the two wave peaks will be the same. This distance is the wavelength (λ). The sound waves travel through the air at a certain speed. If we count the number of peaks which pass a given point during one second (s) we get the frequency of the sound.

The frequency thus gives the number of waves or periods per second. A period is also called a cycle and frequency is sometimes expressed as cycles per second (c/s). One thousand cycles per second is one kilocycle per second (1kc/s). The unit 'cycle per second' is now internationally called Hertz (Hz: 1kc = 1kHz) after a famous physicist of the late 19th century. Each tone corresponds to a certain frequency. The human ear can detect sounds of frequencies from about 20Hz to 20kHz. Sounds made up of tones with frequencies above this range are called supersonic or ultrasonic.

Sound velocity is the distance the sound waves travel per second, which thus corresponds to the frequency multiplied by the wavelength. The sound velocity depends on certain physical properties of the material through which the sound travels, namely elasticity and density. In air, the sound velocity is approximately 322m per second (m/s) at 0°C, and it increases

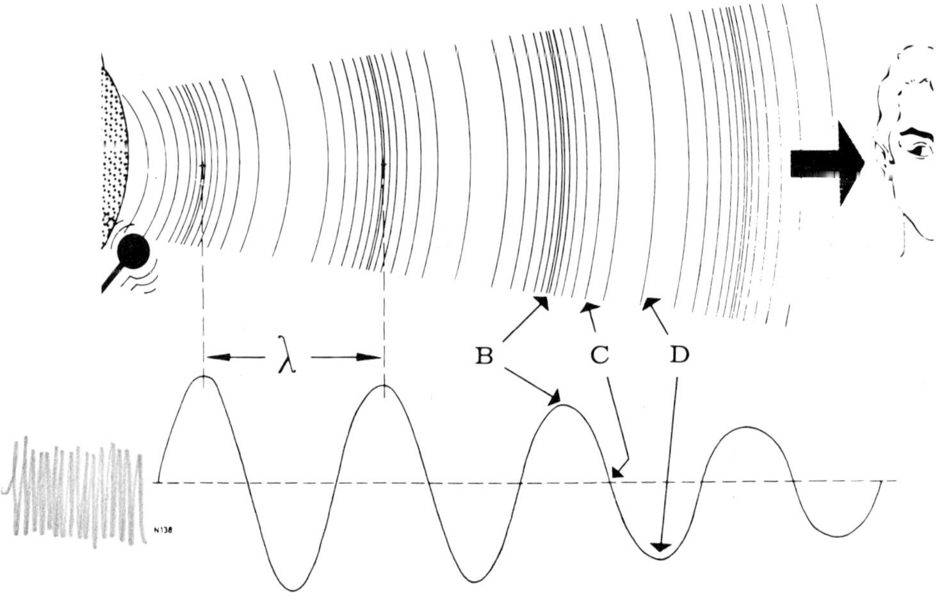

Fig 1 The sound from a gong. The upper part of the figure shows pressure waves, shown below in graphical presentation. λ = wavelength

slightly with higher temperature. In water, sound velocity is about 1,500m/s at 15°C and here also the velocity increases at higher temperatures as well as with increasing salinity. The wavelength of the sound waves can be calculated in a very simple way when we know the sound velocity of the medium and the frequency of the sound. A sound of 15kHz frequency will have in water a wavelength as follows:

$$\text{wavelength} = \text{velocity} \div \text{frequency} = \frac{1{,}500\text{m/s}}{1{,}5000\ \text{Hz}} = 0.1\text{m} = 10\text{cm}$$

Returning again to our gong experiment it is evident that if we walk around a continuously sounding gong we hear the sound the whole time. The sound waves spread in all directions; we say they move regularly from the source of the sound or that the source is omnidirectional. The intensity (or strength) of the sound decreases as we move away from the source. This is because the sound energy is distributed over a rapidly increasing area. *Fig 2* shows that when the distance from the origin of the sound is doubled, the area that it covers is four times greater.

The intensity of the sound decreases at the same rate. Thus if the distance increases 10 times, the intensity per unit area decreases 100 times. Through distribution alone, the sound intensity therefore decreases in proportion with the square of the distance from the source, but there is an additional cause of reduction of sound intensity with distance. Frictional forces in the medium will resist the movement created by the sound waves and sound energy is absorbed. The rate of absorption depends on the frequency of the sound and is greater for higher frequencies.

Sound waves can be directed in the same way as light waves in a light beam or searchlight, and this is done through the form and construction of the source, *ie* the transducer of an echo sounder or sonar (Chapter 2), which is installed in the bottom of the ship. The width of the sound beam is determined by the wavelength of the sound and the size of the transducer surface. With the same wavelength and frequency a larger transducer will give a narrower beam and vice versa. With decreasing wavelength (increasing frequency) one can have a narrower beam using the same size of transducer.

In most acoustic equipment the sound is not given out continuously but in short bursts (pulses) with longer quiet periods in between. The duration of the short burst is called the pulse length. If the duration of the sound pulse is one thousandth of a second (*ie* one millisecond or 1ms), which is usual in many echo sounders, the length of the corresponding train of sound waves in water will be:

$$\frac{1{,}500\text{m/s}}{1{,}000} = 1.5\text{m}$$

Sound is partly reflected when it passes from one medium to another having different density and elasticity. The amount of reflection depends on

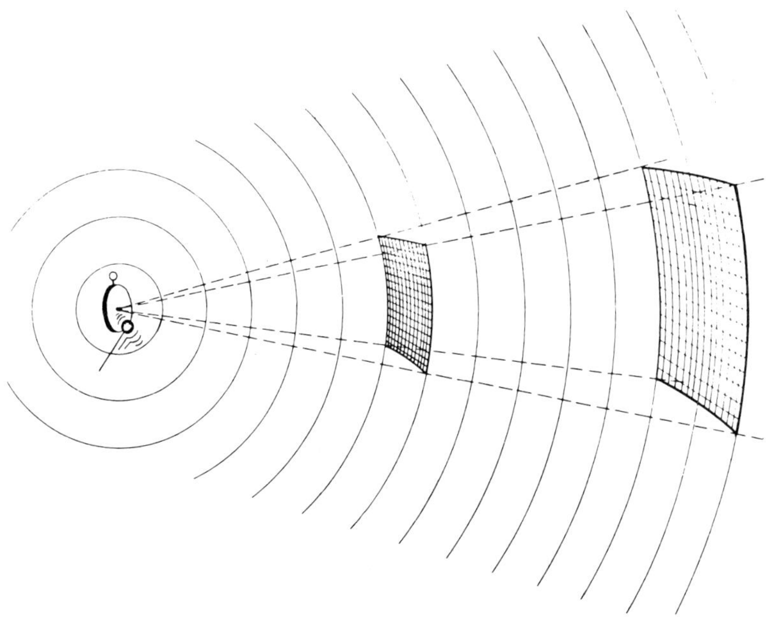

Fig 2 As the sound travels away from the sound source, the area covered by the sound increases four times for each doubling of distance

the difference between these properties of the materials. Air and water are materials of greatly different density and elasticity, and their interface, or the layer at which they meet, is almost a complete barrier to sound. This can be seen in echo sounders when air bubbles are caught under the keel of a pitching or reversing ship, or when passing through the propeller wake of another vessel. The echo sounder is blocked because little or no sound penetrates through the air bubbles into the water. Also, the water surface is an almost perfect reflector for sound, as we shall see when we consider the double echo and sound disturbances below. There is also a great difference in density and elasticity between sea water and rock, which is why rocky ground gives a strong reflection or echoes, whereas the sound reflection from soft, muddy bottoms is less and gives weaker echoes. Fish with a swim bladder give better echoes than fish without, because the difference in density between the water and the gas in the swim bladder is greater than that between water and fish flesh.

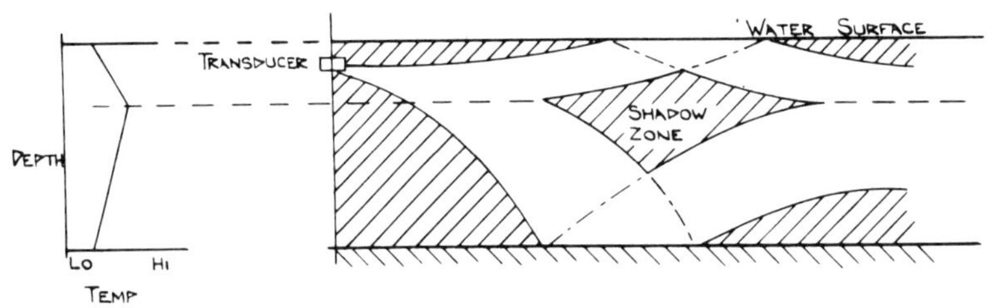

Fig 3 The effect of temperature changes bending a sonar beam. Areas where there is poor reception, or no reception, are shown shaded. Temperature gradient is indicated on the left of the diagram

Within the volume of natural water bodies, refraction, *ie* 'bending' of the sound beam often occurs because of differences in acoustic properties. Regular changes in depth, salinity and temperature affect the propagation of sound if it strikes at an angle of less than 90°. The most significant of these factors is temperature, which produces an increase of approximately 1m/s for a 1°C rise. Consequently, if the temperature increases with depth, a sound beam which is transmitted (from a sonar) in an oblique direction to the temperature layers is bent upwards. Conversely, a decrease in temperature with depth causes such a beam to be bent downwards. The effect is shown diagrammatically in *Fig 3* and illustrates how targets on a sonar display can sometimes disappear. This figure also shows how such a bent sound beam may be reflected by the sea surface or by the sea bed. Vertical soundings from an echo sounder are not affected by refraction.

CHAPTER 2

DESCRIPTION AND USE OF ECHO SOUNDERS

2.1 Main components

An echo sounder consists of four main components (*Fig 4*), the transmitter, the transducer, the receiver and the recorder. The function of the transmitter is to produce energy in the form of pulses of electrical oscillations. In the transducer, the electrical energy is converted to sound in the water and conversely, the sound waves of the returning echoes are converted back to electrical energy. The function of the receiver is to amplify the weak electrical oscillations produced in the transducer by the echo so that they can be recorded or displayed. The echoes can be displayed on a paper

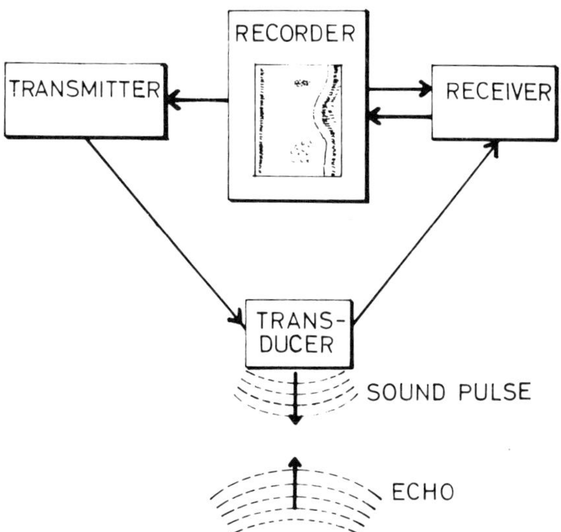

Fig 4 Block diagram of the main components of an echo sounder. A trigger contact in the recorder makes a pulse which starts the transmitter, which produces an electro-magnetic energy pulse for the transducer, where it is converted into a mechanical sound pulse. The echo is received by the transducer and converted back into an electrical signal, which is passed on to the receiver, where it is amplified and fed to the recorder. The recorder also provides the receiver with a start pulse for initial suppression and/or time varied gain (TVG)

recorder, a flash-light display, an oscilloscope (cathode ray tube or CRT) or as an audible signal. The recording unit also has the function of starting the transmitter and of measuring the time between the transmission of the sound pulse and the reception of the echo.

2.1.1 TRANSMITTER

The transmitter generates an electrical pulse at the particular frequency for which the transducer is designed. The block diagram in *Fig 5* shows the main components of the transmitter.

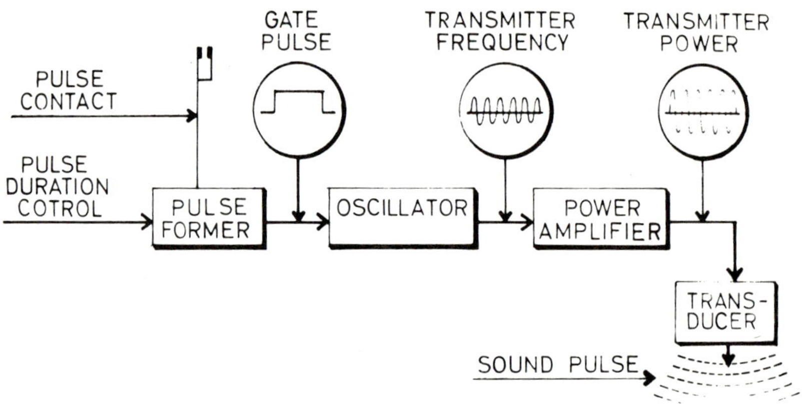

Fig 5 Block diagram showing the functions of the transmitter

A command from the pulse trigger contact in the recording unit tells the pulse former when to function. A gate or switching pulse is produced in the pulse former which determines the length or duration of the pulse. The pulse which is generated by the oscillator is then amplified in the power amplifier before it is fed to the transducer. The power of the pulse produced by the transmitter is an important feature of any echo sounder and it may vary from a few watts (W) up to several kilowatts (kW). It is important to note that the indicated transmitting power of an echo sounder is not always a proper measure of the sound energy transmitted into the water. This is because the efficiency and the directional pattern of transducers may vary considerably.

2.1.2 TRANSDUCER

As already mentioned, a primary function of the transducer is to transform electrical energy into sound energy when sound is to be emitted and conversely, to convert the sound energy into electrical energy when echoes are received.

An additional function of the transducer is to concentrate the sound energy which is emitted into a beam. This beam has the general form of a cone, or of a main beam with some irregularities called secondary beams or side lobes. *Fig 6* shows the directivity pattern of a typical transducer, *ie* the shape of the main beam and its smaller side lobes.

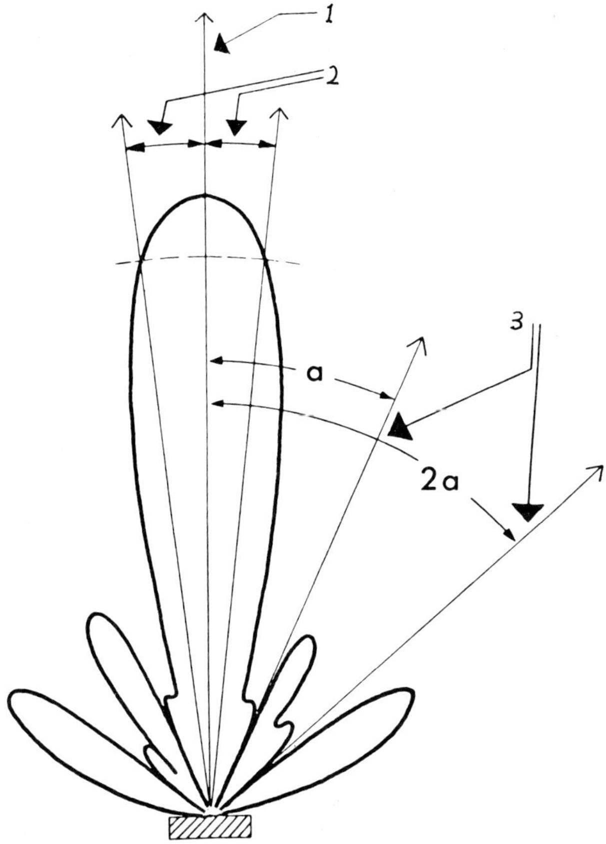

An example of a directivity diagram of a transducer. Key: 1, main beam direction; 2, beam width at the half-power points; 3, directions with minimum power, first minimum shown as 'angle a' and second as 'angle 2a'

From the beam pattern one can determine the so-called beam width, *ie* the angle between the two points on the main beam where the sound intensity is one half the intensity along the acoustic axis (1 in *Fig 6*).

The beam width of a transducer may be estimated with reasonable accuracy by means of the following formula:

$$\text{Beam width in degrees} = 50 \times \frac{\text{wavelength in cm}}{\text{transducer width in cm}}$$

This refers to rectangular transducers. For circular transducers the beam width is approximately:

$$60 \times \frac{\text{wavelength in cm}}{\text{transducer diameter in cm}}$$

There are two main types of transducers for sounding equipment used in fisheries: nickel transducers and ceramic transducers. Their function is based upon the principles of magneto-striction and electro-striction respectively.

The quality of transducers may vary considerably, and is measured by their efficiency, *ie* the ability to transform electrical energy into sound waves and vice versa. It is also generally desirable that the secondary side lobes produced by a transducer should be as small as possible, not only because they represent loss of energy, but also because they may cause unwanted echoes which can confuse the interpretation of the echo traces. (For exception see *Fig 20, 11.1* and *11.6*). *Fig 7a* and *Fig 7b* show different types and sizes of ceramic and nickel transducers. (The ceramic transducers are on the left of *Fig 7a.*)

2.1.3 Receiver

The weak electrical signals produced in the transducer when echoes are received must be amplified several thousand times before being passed on to

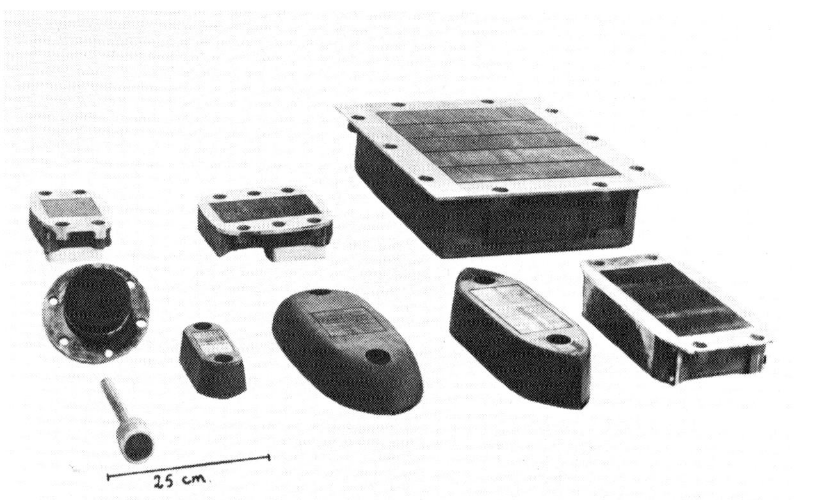

Fig 7a Different types and sizes of transducers, with and without streamlined housings. The beam width of a transducer can be estimated from its size and its working frequency

the recorder. This amplification takes place in the receiver and the amount of amplification can be varied by the sensitivity or volume control. In order to reduce or remove the echoes from very close range which are much stronger than those from a distance, the sensitivity of the receiver is automatically suppressed so that it is low at the moment of transmission but

Fig 7b Installation of a 15cm x 30cm 38kHz magneto-striction transducer. A clean face is very important for the proper function of a transducer

increases with time to full sensitivity. The extent of this initial suppression of the sensitivity varies somewhat from one type of echo sounder to another but usually it is designed to affect echoes from targets within 10 to 20m of the transducer.

In certain modern echo sounders, the automatic sensitivity control is adjusted to cover the complete operational depth range of the echo sounder and to compensate for dispersion and absorption of sound within this range. The strength of a returning echo is thus independent of the depth from which it comes and dependent only on the echo target strength of the reflecting objects, such as fish or bottom. This type of depth (or rather distance) adjusted amplification is known as Time Varied Gain (TVG) and will be described further under *4.4*.

2.1.4 Recorders

We will first consider an ordinary paper recorder.

Various types of paper recorders are used, but the principles of their function and their main components are generally the same. *Fig 8* shows diagrammatically these main components: the motor, the gearbox, the stylus or marking pen, the paper, the paper drive, the transmission contact and the suppression contact.

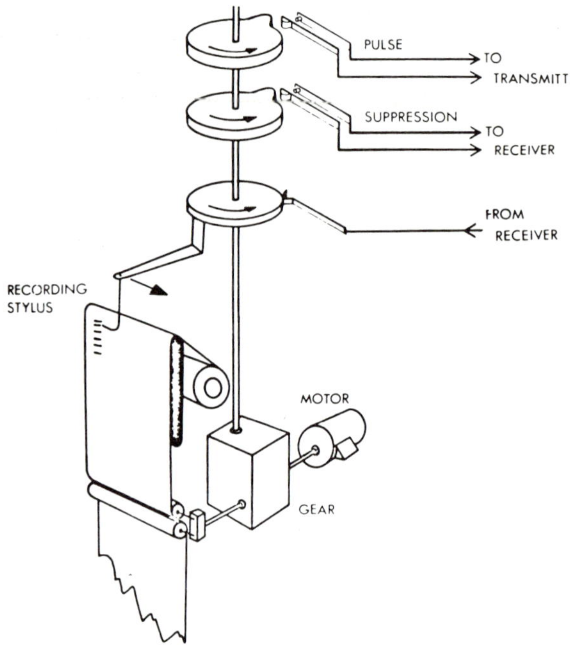

Fig 8 Diagram showing the function of a rotating pen recorder

Figs 9, 10 and *13* show actual chart and stylus systems. The oldest recorder system (*Fig 13*) with a rotating arm or disc is still used in combination flasher/recorders. It has the disadvantage that the pen movement is not straight but is circular, so producing a curved trace on the echogram.

Besides recording the echoes on paper, the recorder performs several coordinating functions. It gives the signal for triggering the transmission pulse, and at the same time for the initial suppression of reception operated each time the stylus passes the zero mark on the depth scale.

The recorder also measures the time gaps between transmission of the sound pulses and reception of the echoes. This is done by moving the recording stylus over the paper at a certain uniform speed which is dependent on the sound velocity and the depth scale at which the echo sounder is operated. The speed of the stylus at various depth scales is usually adjusted to the 'normal' speed of sound in the sea, *ie* 1,500m/s. More sophisticated echo sounders have provisions to conveniently adjust the stylus speed to various sound speeds (fresh or sea water; varying temperature or salinity). Since the sound must cover the distance twice—down and up—the stylus speed must correspond to half the speed of sound. For instance at a depth scale of 750m, the stylus would pass from 0 to 750m in one second.

The paper speed is usually adjusted so that the lines recorded by the stylus are closely spaced. There are echo sounders where the paper speed can be varied. The appearance of fish and sea bottom presented on the recording paper will depend on the stylus speed, the paper speed and the speed of the ship over fish and ground (*4.1*).

The recorder usually has several depth ranges, for instance 0–125m and 0–250m. A change of depth range means a change of stylus speed which is achieved by changing the gear ratio or by changing the speed of the motor. The basic depth ranges may sometimes be extended to cover greater depths. This is done by a simple adjustment of the timing of transmission in relation to the position of the stylus. For instance, using a basic depth range of 0–125m one can select a depth interval of 100–225m or 200–325m, *etc*. This is sometimes referred to as a depth scale of 0–125 phasing to 225 or

Fig 9 Straight line recorder, with stylus arm rotating in a rather complicated curvature

325m, or a phased display. This method of choosing a certain depth interval with high stylus and paper speed is an advantage because it improves the resolution of the display which can, for instance, help to distinguish fish echoes from bottom echoes.

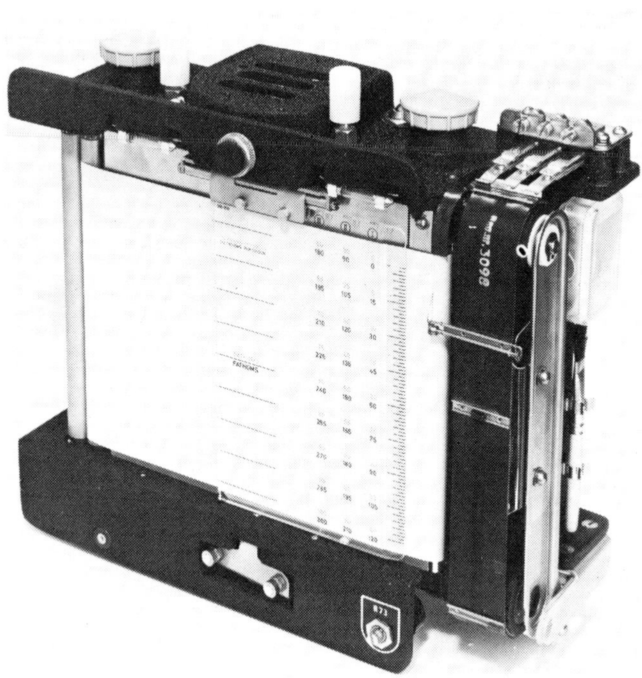

Fig 10 Recorder with belt driven stylus

There are two different kinds of recording paper in use for echo sounders, wet paper and dry paper. The wet paper contains a chemical solution which is coloured black/brown when an electric current is passed through it. The amount of colouring is proportional to the current strength which in turn is made proportional to the strength of the echo, from either the bottom or from fish. Wet paper has a high dynamic range, *ie* the number of differences in the density of the colouring which can be detected is high. It therefore provides good resolution of echo or target strength.

However, care must be taken when storing unused wet paper so that its humidity, and thereby its sensitivity, is not reduced. Recordings made on wet paper will not keep. On dry paper, the echo traces are produced by electric sparks from the stylus burning off a light grey surface layer, and this can be stored without risk so that the echograms will keep. Dry paper has a smaller dynamic range for recording echo strength than has wet paper.

The multi-stylus type of recorder has no moving stylus, but instead has several hundred stationary styli mounted as a comb across the paper. The echo signals are fed on to the styli electronically and the time between the transmission and the return of the echo is measured by an electronic time generator. This type of recorder has the advantage that, apart from the paper drive, there are no moving mechanical parts. *Fig 11* shows diagrammatically the main components of a multi-stylus recorder and *Fig 12* shows the display and comb.

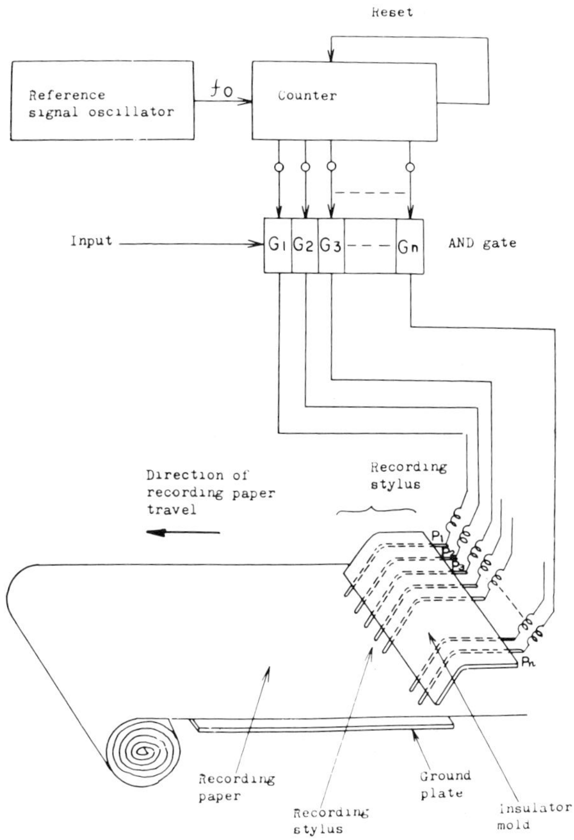

Fig 11 A multi-stylus recorder unit, shown schematically

When the information required from the echo sounder is very simple, to indicate only the water depth for example, a simple flash-light indication may be sufficient. In such a system, a small lamp is rotated over a depth scale and when the echo is received from the bottom, this causes the lamp to flash at the appropriate point on the scale. No recording paper is needed by this method, but the information presented is not stored and must be watched as required. Recorders combining paper display with flash indication are also available. With these it is possible to have both types of display at the same time or to select 'flash only' and thus save paper (*Fig 13*). In

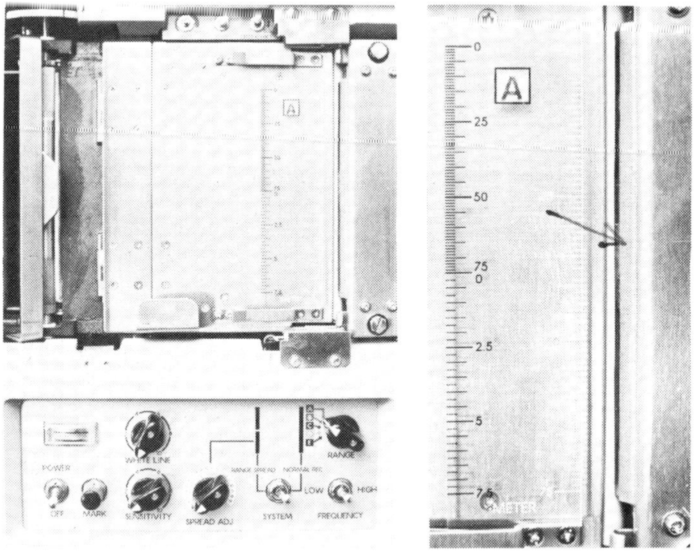

Fig 12 *(Left)*, Multi-stylus recorder. *(Right)*, Multi-stylus comb

Fig 13 Recorder combined with flash-light display. The echogram produced by this type of 'windmill' recorder has a curved trace

order to exclude misinterpretations, flasher echo sounders are usually adjusted so that they indicate only the strong bottom echo and not fish echoes. To use the flash display for fish finding requires experience, or specific technology, in order to distinguish between different echo strengths.

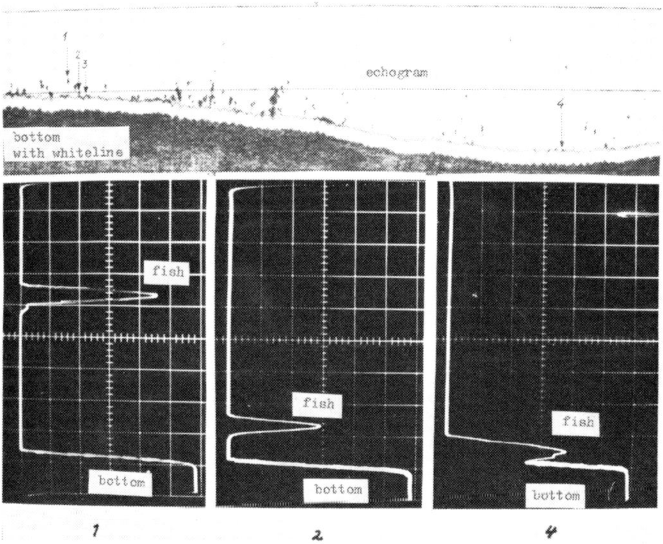

Fig 14 Echogram of fish close to the bottom (above). Fish marked 1, 2 and 4 are shown below on an oscilloscope with bottom lock. For whiteline see *2.2.3*

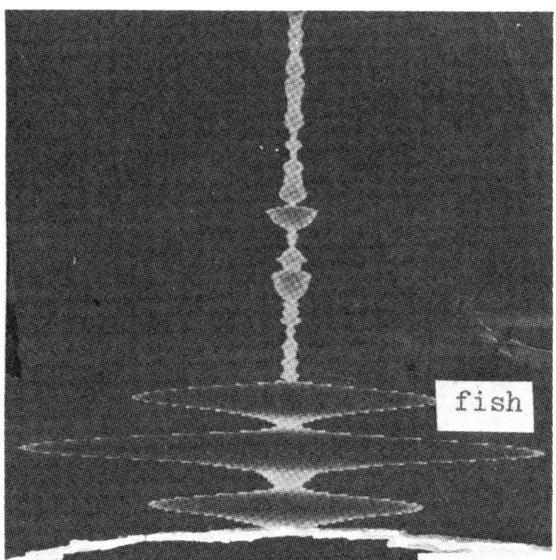

Fig 15 Photograph of the CRT display of a commercial fishlupe, in 5fm range

Oscilloscope or cathode ray tube (CRT) presentation of echo signals is sometimes known as fishlupe. The depth range to be studied and the scale of display can easily be varied in an oscilloscope to provide optimum resolution and it also provides true echo strength indication. This display is therefore the best for the detailed study and identification of echoes, while the CRT display is well suited to the recognition of fish echoes close to the bottom. This feature can further be improved with a bottom lock system (*4.3*). *Fig 14* shows one example of how this system works and *Fig 15* is a photograph of a display on a commercial fishlupe. To achieve superior overall performance it was an obvious step to combine paper recording with CRT display and so conveniently utilize the advantages of both systems. Such units naturally belong to the more powerful and expensive class of echo sounders (*Fig 16*).

Fig 16 Example of a fish-finding echo sounder which combines paper recording and CRT display

2.2 Controls

2.2.1 On-off switch

All types of echo sounders have a main on-off switch. For smaller and simpler types of echo sounders this switch will also start the instrument functioning and may further include the sensitivity control. In other types, the on-off switch is used only to put the instrument ready for use and a special switch starts the sound emission and the recorder. This arrangement allows for the electronic components of the echo sounder to be heated up. One should note that electronic components are not generally worn out by being kept ready for use; indeed their life may be reduced by frequent on-off switching.

2.2.2 Sensitivity

The sensitivity of the receiving system is only one of the factors which determine the maximum depth ranges for various targets. Other factors of importance which may vary from one echo sounder to another are the frequency, the size of the transducer and the transmitting power. The reflecting property (target strength) of the fish is also of importance. All these various factors will be discussed elsewhere in this manual. Here we will only point to some main rules concerning appropriate sensitivity settings.

Because of increased dispersion and absorption of sound with depth, the echoes received from greater depths will be much weaker than echoes from targets close to the transducer. It is therefore important to adjust the sensitivity to the depth which is being surveyed. It is not possible to give any general rules as to which sensitivity setting to use at different depths. The sensitivity scales vary widely between different types of sounders, but very often fishermen tend to use too low sensitivity when searching at greater depths. The reason for the reluctance to use high sensitivity for deeper waters may be the experience that plankton and other small organisms in the surface layers may show up as very heavy traces. Such surface layer traces can be misleading, particularly when searching for near-the-surface fish during the night, but they can be ignored in deep water search.

The upper limit of any sensitivity setting is that which gives too strong a display of background noise. Therefore this noise must be reduced as much as possible, otherwise fish traces may be obscured and become difficult to identify. The best way for the fisherman to get acquainted with his echo sounder and its sensitivity settings is to make a record of the gain settings used at different depths and for different types of fish. This may then be developed into a personal code of practice for the operation of his particular echo sounder for its various purposes.

2.2.3 WHITELINE AND GREYLINE DISPLAY

The purpose of the 'whiteline', 'greyline' or similar methods of chart display is to extend the dynamic range recording of the system which, as already mentioned, is limited both in wet and dry paper. This is done by a special signal treatment in the receiver whereby echoes above a certain strength either completely block the receiver (whiteline) or are presented as a faint grey recording (greyline). This type of signal treatment has proved particularly useful in attempting to distinguish traces of fish close to the bottom from the bottom trace itself. Even though fish echoes are considerably weaker than bottom echoes, they may still be strong enough to produce dense marking on the echogram. When the fish are close to the bottom, the two traces will merge and the fish trace will be indistinguishable from that of the bottom. The whiteline or greyline effect will present the strong bottom echo as a white or greyish trace so that the black fish traces close to the bottom can better be identified (see *Fig 14*, upper section).

The echo level at which this special signal treatment occurs can be varied in some echo sounders. In addition to the separation of fish traces from bottom echoes the arrangement can also be used to assess the density of fish schools and to indicate the nature of the sea bottom, *ie* whether it is hard or soft (Chapter 6).

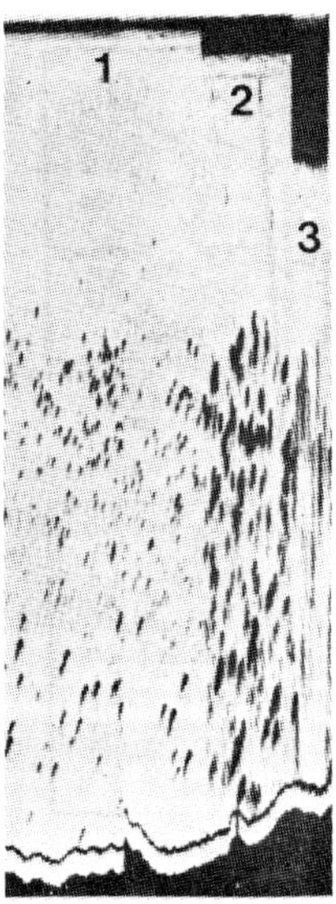

Fig 17 Echogram showing the effect of different pulse lengths on single fish. Area 1, 1ms; Area 2, 3ms and Area 3, 12ms

2.2.4 POWER AND PULSE LENGTH

The pulse power of most echo sounders lies between less than 100watts (W) to about 1 000watts (1kW). Special echo sounders for deep sea work and other purposes may have a pulse power of 10kW or more, however.

On echo sounders with variable pulse power, it is generally advisable to use a high pulse power and low receiver gain in order to obtain clearer and better noise-free recordings. Only when it is advantageous to reduce signal strength, for example, when sounding at short range or in shallow water, should the low pulse power be applied.

The pulse length is the duration of the sound emission from the transducer. A pulse length of about one millisecond (1ms) is normally used in medium frequency instruments. For high frequencies the pulse length may be as short as 0.1ms. Short pulses will increase resolution and thus make it easier to distinguish echoes from fish which are close to the bottom and echoes from individual fish (*Fig 17*), also to judge the density of fish schools.

2.2.5 Depth range

To obtain a maximum of information from an echo sounder, it should be able to survey the widest possible depth range, but at the same time provide detailed information about any fish targets within the water column. These requirements are partly contradictory. However, they can be met, either by alternately switching between the full maximum depth range and a phased

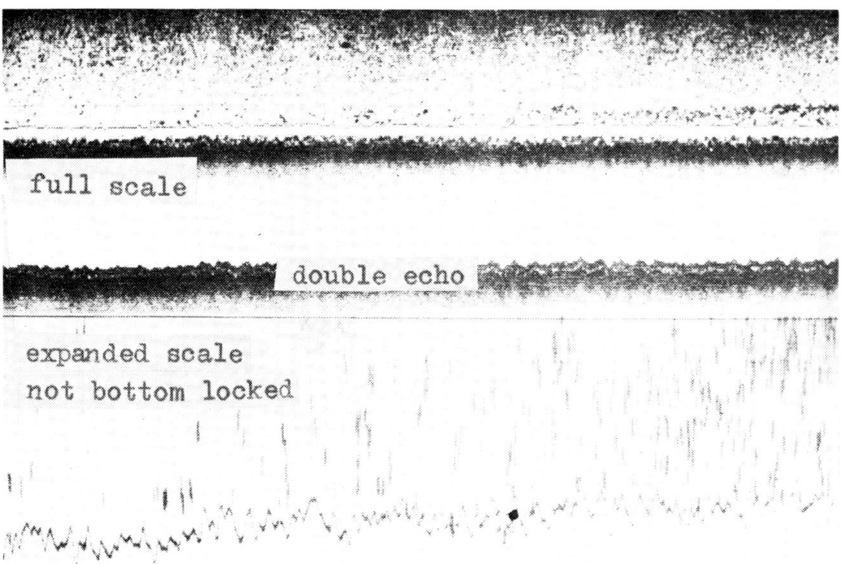

Fig 18 Echogram from a multi-stylus recorder, the upper part showing midwater (pelagic) fish and fish close to the bottom. In the lower, expanded part, the near-bottom fish are shown in more detail

shorter range with higher pen speed on the same recorder or by using an additional oscilloscope, with a smaller range that can be shifted to any depth.

With a multi-stylus recorder or an electronic scale expander (see *4.3*), a wide survey-range can be shown simultaneously with a more detailed expanded scale on the same width of paper (*Fig 18*) although the use of the same paper speed for the two depth scales slightly limits the usefulness of this arrangement.

An example of switching stylus speed on the same recorder is shown in *Fig 19*, and in the left hand section of this echogram a layer is shown above the sea bed. For a closer study of this trace, the echo sounder was switched to a smaller, phased range with a faster stylus and paper speed. With the resulting expansion of the layer, it can be seen that it is not a group of small organisms but consists of fish-like traces rising to about 10fm above the sea bed. These traces were identified as small squid.

However, it is not always preferable to operate at a small range with high stylus and paper speed, particularly when trying to obtain information about

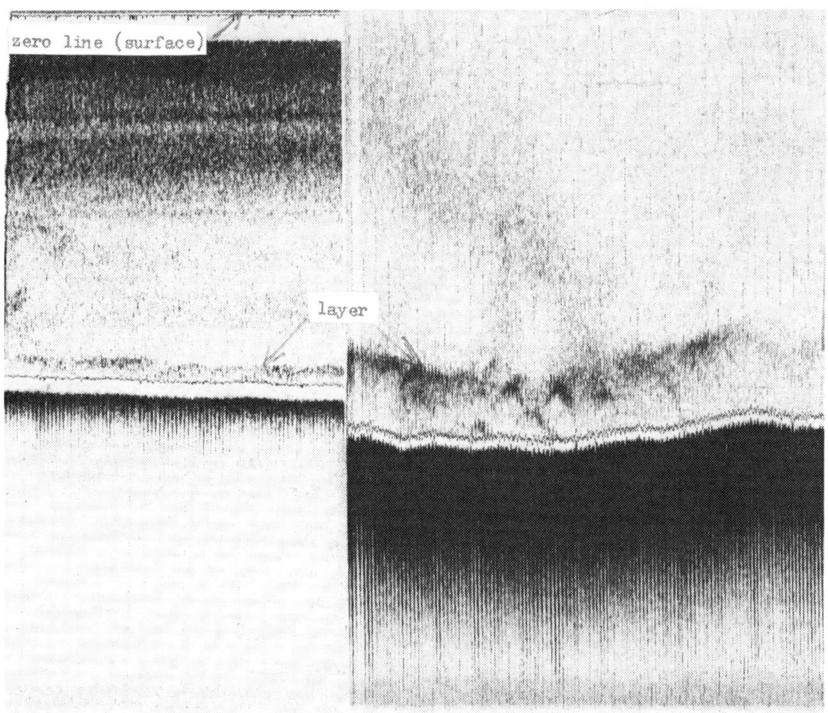

Fig 19 An example of range-switching, in order to better identify a layer trace. Note that the smaller phased range on the right does not include the zero sound and the upper water column

the sea bed. For example, the left hand section of the echogram in *Fig 20*, with a paper speed of 4mm/min, clearly shows all the characteristics of transition from soft to hard ground—wider whiteline, darkening and increase of the tails (echo length) of the bottom trace, and darkening of the second bottom trace with occasional whitelining. It also shows more clearly the side lobe traces which also indicate hard ground. They are displayed to the full length as oblique stripes between first and second bottom echo, thus giving additional warning of hard ground.

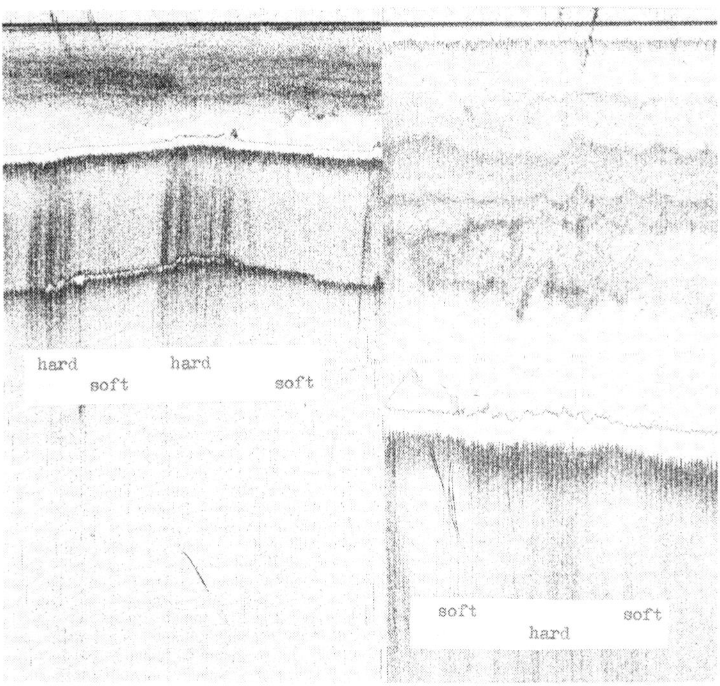

Fig 20 Range-switching may lose some hard bottom detail

With the smaller range recording (right), the second echo is not shown, so that related indications of hard bottom are lost. With the higher paper speed (12mm/min) and higher stylus speed, the side lobe traces are less noticeable also, and only the increase in echo length and the widening of the whiteline of the true bottom trace remain as a warning to beware of hard bottom.

2.2.6 Paper speed

A change of the range displayed across the paper means altering the speed of the stylus; the faster the stylus speed, the shorter the range displayed. At the same time the paper speed must be adjusted either to prevent overprinting lines or having gaps between lines and thus breaking the echogram. Commercial echo sounders provide for this. Some also offer a choice between two basic paper speeds, providing paper saving or better resolution as desired.

2.2.7 Stylus speed

Since the accuracy of the displayed depth depends upon the time taken for the stylus to travel across the paper, the stylus speed should be checked regularly to ensure that it complies with the value given by the manufacturer. This can be done with a stop watch, making the check some 5 minutes or more after switch-on to allow time for the unit to warm up. In some units the speed can be checked by means of a calibration switch which, when operated, produces a number of uniform markings across the paper. The speed can then be corrected to give the required number of marks per scan. Some special sounders used for great depth, for different water temperatures and salinity, or for specific accuracy requirements, have an indicator graduated in sound velocity.

CHAPTER 3

INSTALLATION AND MAINTENANCE OF ECHO SOUNDERS

3.1 Positioning and installation of the transducer

The positioning and installation of the transducer are often decisive in ensuring proper performance. Normally the transducer should be placed between one-third and one-half of the vessel length from the bow and approximately one metre away from the keel. In vessels smaller than 30m the transducer may be closer to the keel.

The ideal position of the transducer is at the deepest, and at the same time most noise-free point of the vessel. This point may be difficult to determine since the propeller noise and the noise produced by the flow of water along the hull may vary with the trim and the speed of the vessel. The deepest point is usually by the propeller or the rudder, but as these are the most important sources of noise, the transducer must be placed well away from them (forward). Neither should the transducer be too close to the bow, for this is not only a source of noise, particularly at full speed, but its vertical movements also interfere with the quality of the recordings. In bad weather, recordings may be interrupted by the highly aerated water thrown under the ship in this way and this is an additional reason for avoiding positions too close to the bow.

The transducer is usually housed in a streamlined 'shoe' or blister to reduce water noise and turbulence (see *Fig 7*), and *Fig 21* shows how the transducer should be mounted in relation to the keel. The angle of between 5° and 8° will vary according to the form of the bow and the distance from it. The surface of the transducer should always be tilted about 2° forward (see *Fig 21*) in order to prevent turbulence on the transducer itself while underway. This tilting also has the effect of 'shadowing' the transducer face from the propeller noise.

An alternative method, described by the Japanese, proposes transducers with a brim (*Fig 22*). The transducer is mounted about two-thirds of the vessel's length from the bow and projects below the hull in order, if possible, to avoid the air bubble layer reverberation (*3.4*). For a 30–40m vessel, the projection should be greater than 80cm.

The brim is claimed to operate as shown in *Fig 23* and to reduce considerably the attenuation due to the air bubble layer (see *Fig 24*). Probably because of structural and operational problems, these projecting echo sounder transducers, in spite of their undoubted advantage for reducing attenuation, have not found ready acceptance in fisheries or shipping.

As a rule, the instrument producers and their representatives have experience in transducer installation and advantage should be taken of this as far as possible, for each individual case.

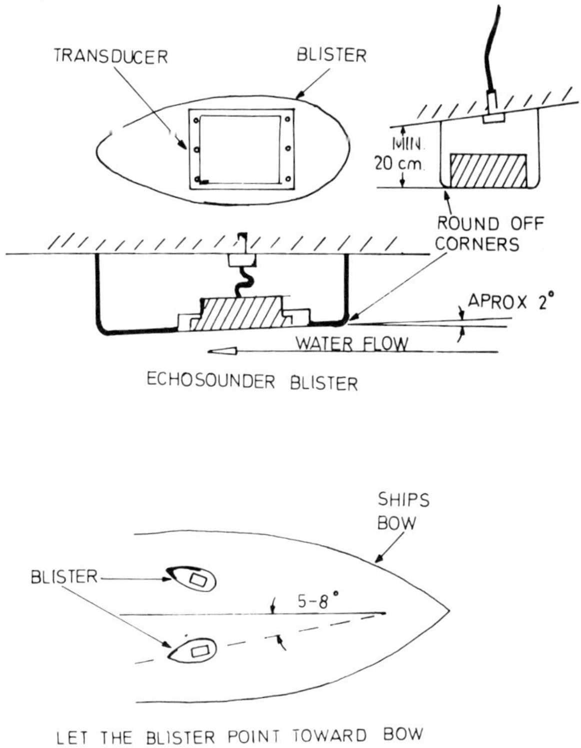

Fig 21 Example of transducer installation. The top sketch shows detail of the transducer mounted inside a streamlined blister. The lower sketch shows the positioning of blisters

3.2 Installation of the recording unit

The recording unit should be installed so that the display can clearly be seen and the controls conveniently operated from the skipper's position when the vessel is fishing, and also when under way.

3.3 Power source and electrical installation

Perhaps the most common and widespread cause of faulty or poorly functioning echo sounders is an unsuitable or failing electrical installation. The installation must ensure sufficient power for the transmitter, correct voltage for the receiver and, in addition, be as free as possible from all electrical noise. The problem of correct voltage is particularly important on small vessels where 12V and 24V batteries comprise the power source. The proper installation and maintenance of the general electrical equipment is thus a fundamental requirement for obtaining satisfactory performance from an echo sounder.

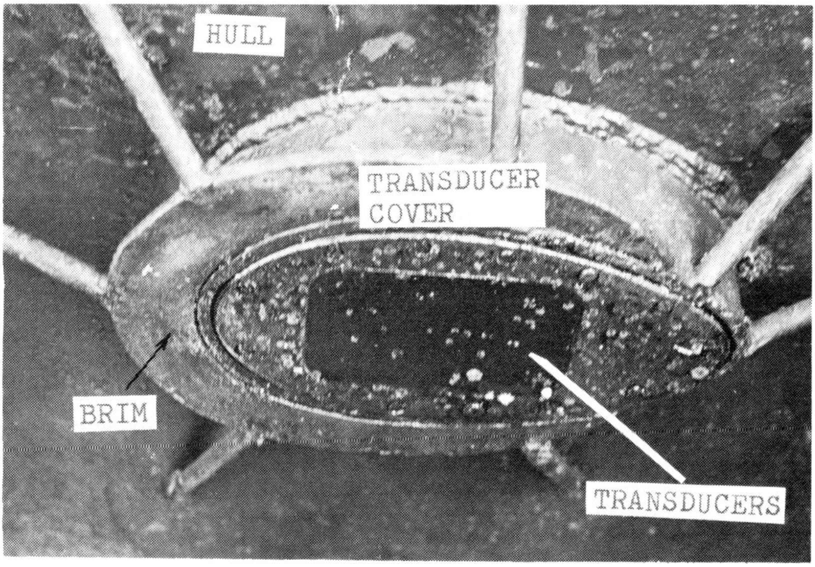

Fig 22 Streamlined transducer cover with a brim

3.4 Noise

Noise may considerably reduce the efficiency of echo sounding and it is therefore of the greatest importance to avoid or reduce noise as far as possible. The types of noise can be classified as follows:

Self-generated noise	**Reverberation**	**Ambient**
Hydrodynamic	Volume	Sea
Machinery	Bottom	Biological
Propeller	Surface	Man-made

Noise from outside sources is generally of less importance than self-generated noise. When many vessels are working in a limited area their propellers and acoustic instruments may disturb each other's echo sounders. Big waves may produce acoustic noise and whales or dolphins can emit sound signals which are recorded by echo sounders. Fishing gear such as bottom trawls generate noise, but this is only of importance for acoustic instruments mounted on the trawl itself (netsonde, *5.4*).

The reflection of all kinds of sound from plankton and air bubbles in the water volume and from the boundaries of the medium, *ie* bottom and surface, is often classified as noise under the term 'reverberation'. This is further discussed under *8.2*. When particularly heavy, this reverberation or interference can obscure echoes from fish and even the bottom (*Fig 25*).

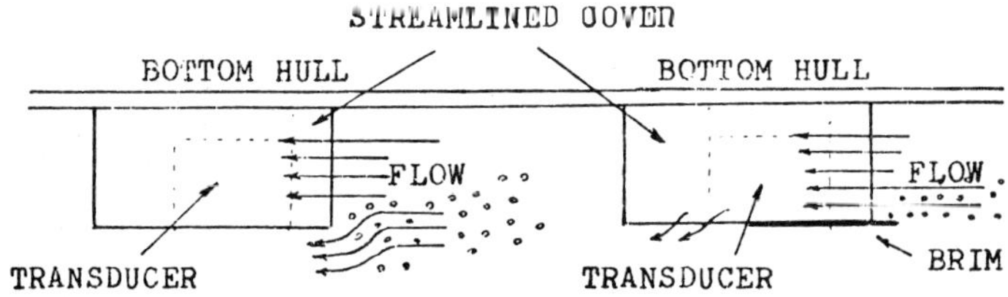

Fig 23 Effect of a brim fixed to the protruding cover of the transducers against water flow which contains air bubbles

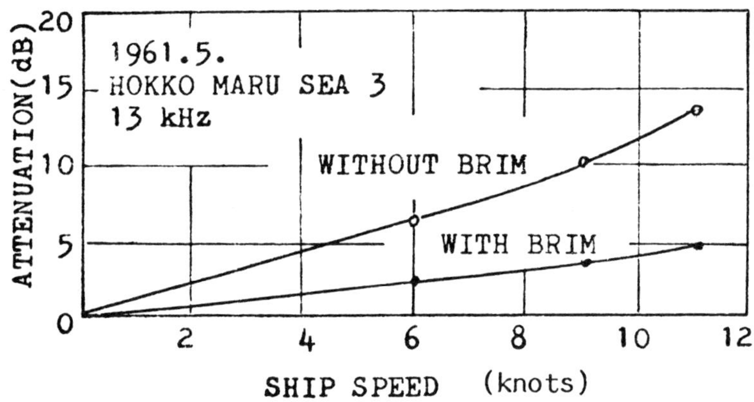

Fig 24 Measured data showing how a brim can reduce attenuation. This is seen to be lower at all speeds when a brim is attached

A reduction in the gain will normally make it possible to distinguish the echoes coming from the bottom from those caused by fish, if they are strong enough. In echo sounders with time varied gain (TVG), this problem is less pronounced, except for longer sounding distances. As already mentioned (*3.1*) it is also possible to avoid the bubble layer by projecting the transducer below the bottom of the vessel (*Fig 26*). The attenuation caused by air bubble layers is less with sounders of higher frequency.

Self-generated noise is derived from various sources in the vessel and its movement through the water and often limits the performance of echo sounders. Vessel noise may originate from the propeller and sterngear, (cavitation, vibrating blades or shaft) or from the many parts of moving machinery (main and auxiliary engines, reduction gears, pumps, ventilators). The most serious noise problems are not always created by what may appear to be the most obvious main sources; a small amount of damage to a propeller blade or piece of machinery working at a particular frequency may seriously affect acoustic instruments. Another important source of

Fig 25 An example of how noise appears on the recording. On this echogram, it is difficult to distinguish the bottom at a depth of 1,000m

vessel noise is that generated by the water passing over irregularities of the hull. This will increase greatly with speed. Occasionally the vibration of loose or poorly mounted objects may be a source of noise.

The position of zinc anodes may affect the echo sounder's performance. They should not be placed forward of the transducer but, if this is necessary, one must avoid a position which may result in aerated water passing the transducer when the vessel is travelling at speed. *Fig 27* shows how zinc anodes should be placed near the propeller and rudder in positions where the water flow past the propeller is lowest.

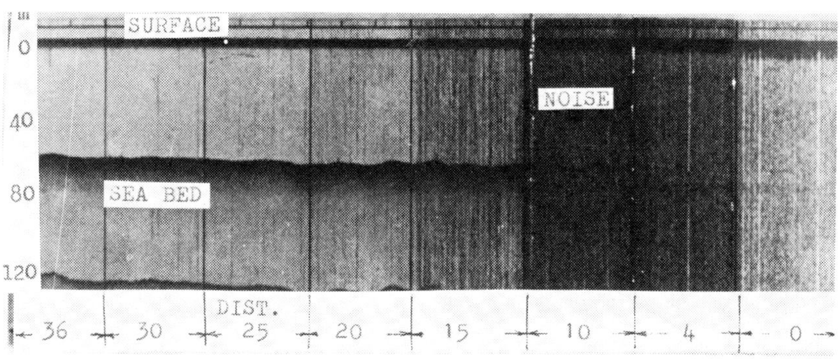

Fig 26 Result of an experiment to show the relationship between the appearance of reverberation and the distance which the transducer projects below the bottom of a 20m boat ('DIST' on echogram illustrated)

Since most of these self-generated noises increase with the ship's speed the simplest means of reducing them is to reduce speed during periods of careful searching. If possible, the transducer should be mounted in such a manner that the propeller cannot be seen from the face of the transducer, and it is remote or isolated from the engine room machinery.

Electrical noise from such sources as generators, electric motors or relays may likewise cause serious disturbances. Some of this type of interference can be avoided by screening certain cables, and by correct earthing.

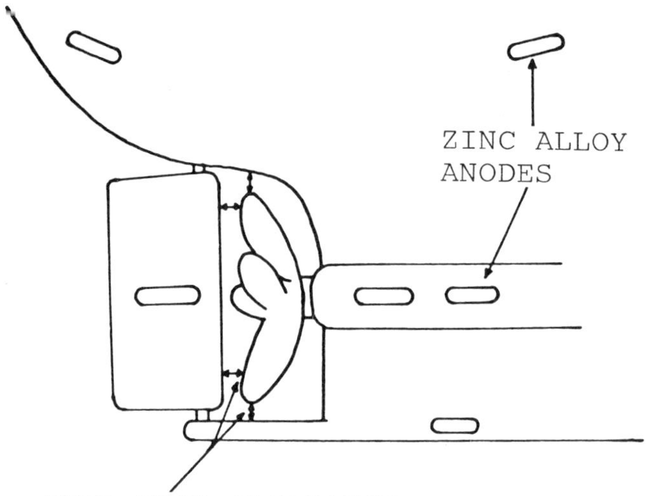

Fig 27 Recommended propeller arrangement and positioning of zinc anodes to reduce noise

3.5 Maintenance

The transmitter and receiver do not usually require any routine maintenance by the operator, but paper recorders should be looked after carefully. Whenever possible the transducer should be inspected and its surface cleaned.

The transducer face must be kept free of growth in order to function properly. Scratching the surface with a hard sharp instrument or abrasive material will damage it and soft implements such as pieces of wood should be used. The transducer face should not be painted in any way.

The maintenance of the paper recorder should strictly follow the advice set out in the instruction manual for the make of instrument installed. It should also be remembered that one drop of oil too much may do more harm than one too little. Dry paper recorders should be cleaned regularly to remove the carbon dust produced from marking the paper, and the stylus (or styli) should be inspected, cleaned, lengthened or replaced regularly.

CHAPTER 4

MORE ABOUT THE FUNCTION OF ECHO SOUNDERS

4.1 Resolution

Resolution is the ability of the equipment to identify and distinguish between targets, *eg* two fish or fish and sea bed. The resolution capability of an echo sounder is dependent on the following factors:

— wavelength (or frequency) of the sound
— pulse length and form
— width of the sound beam
— speed and range of display

We have already seen that the wavelength is equal to the sound speed divided by the frequency (Chapter 1) and that in practice echo sounders have pulse lengths in the order of lms or 1.5m. For better resolution the pulse length should be reduced but there is a relationship between the minimum effective pulse length that can be produced and the sound frequency used. In order to emit a maximum of sound energy, the transducer must receive a certain minimum number of cycles of electrical energy. A good arrangement of transmitter and transducer will allow a maximum sound emission after three or four cycles. The time duration of an effective pulse can therefore be less for high frequencies. For example, a sound frequency of 100kHz corresponds to a wave length of 1.5cm and a sound emission of 3 to 4 cycles will be 4.5 to 6.0cm in water. Whereas at a lower frequency of, for example, 15kHz, maximum sound production will require a much longer time and result in a sound emission of 30 to 40cm. Since the transducer also needs some time to return to an inert state after excitation, the minimum effective pulse length will be approximately twice the figures indicated in the examples.

It is of interest to know what detail can usually be distinguished by an echo sounder. A simple example of the vertical resolution is shown in *Fig 28*. It is evident that if the difference in distance of two targets, such as fish, from the transducer is less than half the pulse length their echoes will converge. The pulse length in this example is 1.5m corresponding to 1ms. The fish therefore need to be separated by more than 75cm in order to be distinguished as two separate echoes. Because of inertia in the transducer, recorder, *etc*, the distance is in practice somewhat greater than half the pulse length. For good resolution it is important to have a short pulse, low inertia of transducer and electronic units, and high recorder pen speed or oscilloscope sweep.

A similar problem exists when resolving fish echoes horizontally. Echoes from fish at approximately the same distance from the transducer can produce a trace similar to that from a single fish. This effect can be reduced by narrowing the beam (see *Fig 29*) and this is especially important when

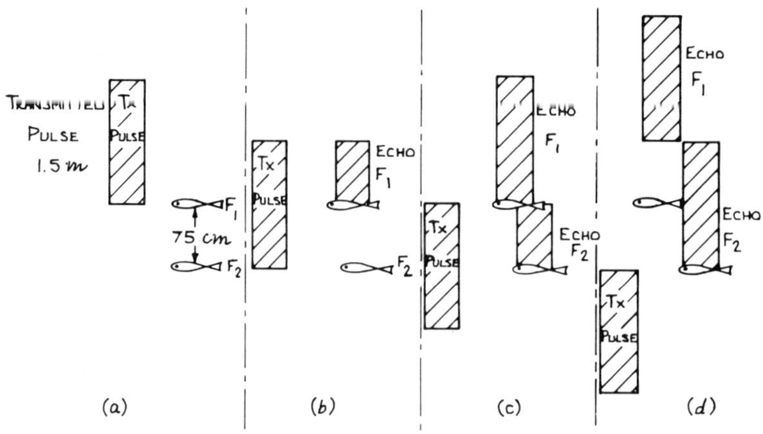

Fig 28 Simplified diagram of echo from two fish separated by half the length of the sound pulse

operating at greater depths because of the beam spreading. There is, however, a limit to how narrow the beam can be made since, with extreme roll of the ship and at high speeds, it could be possible to pass fish between soundings. On sloping or irregular bottom, depth indications could also become confusing. For this reason, when very narrow beams are required, the transducer must be stabilized so that it always points straight down to the bottom.

The paper speed of the recorder and the speed of the vessel also influence the amount of detail in, and the character of traces. Examples are illustrated in *Figs 30* and *31*.

In *Fig 31*, Section 1, the trawl is being shot and in Section 2 the vessel is trawling at a speed of 3 knots over what appears to be difficult ground. The speed of the paper is 2.5mm per minute which, at 3 knots, represents 92.6m *ie* 37m for lmm of paper. In the vertical direction, the paper scale is lmm per 3.35m. There is therefore an exaggeration of the vertical features by 11 times in relation to the horizontal. The true variation in sea bed is a slow change of depth over a distance of 1km and therefore the ground is suitable for trawling. In Section 4, the same ground is traversed, but at a speed of 9 knots. The paper speed and range settings are the same but the vertical features are now exaggerated 33 times in relation to the horizontal. The user should therefore establish the scaling effect for different range scales and paper speeds in order to obtain the correct impression of variations in the bottom contours.

4.2 Recording of fish close to the bottom

Even with the improved signal treatment of whiteline or greyline effect (*2.2.3*) there still remain problems connected with detecting fish close to the bottom. Because of the width of the sound beam, part of the echo from fish close to the bottom will be obscured by the stronger bottom echoes returning from the central part of the bottom which is closer to the transducer. The

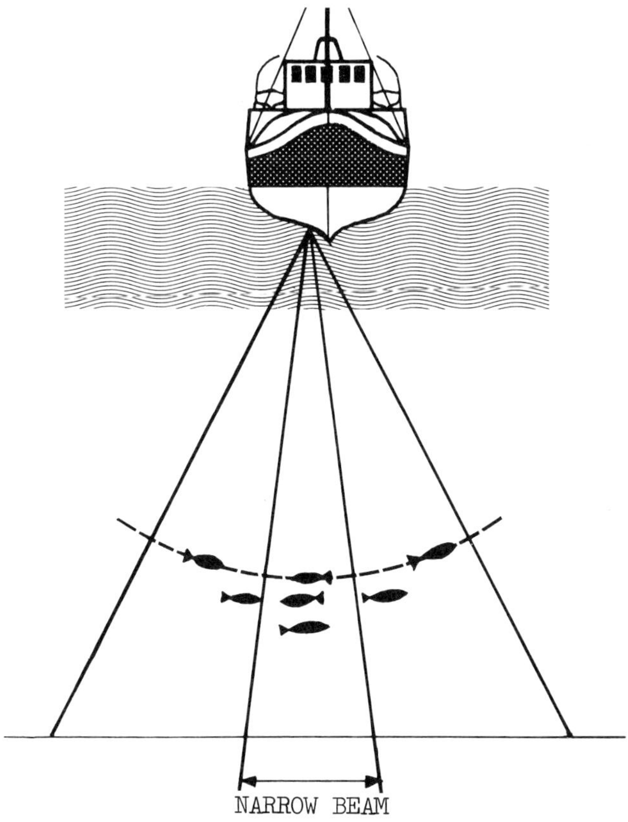

Fig 29 Improved horizontal resolution of fish, obtained by using a narrow beam

field in which fish echoes are obscured is called the 'dead field' and *Fig 32* illustrates this phenomenon.

When the bottom is uneven or is sloping steeply the dead field will increase. The extent of the effect of the dead field can be seen in the following example.

With a normal beam transducer having a 12° beam angle and an even bottom at a depth of 200m, approximately half of all fish in the beam within 2m of the bottom will be in the dead field and thus concealed by the bottom trace. In this context it should be noted that many small bottom trawls have a headline height of approximately 2m.

The resolution of the echo sounder is of importance in determining how close to the bottom fish can be detected. A narrow beam width reduces the dead field effect so that fish quantities recorded off and near the bottom are

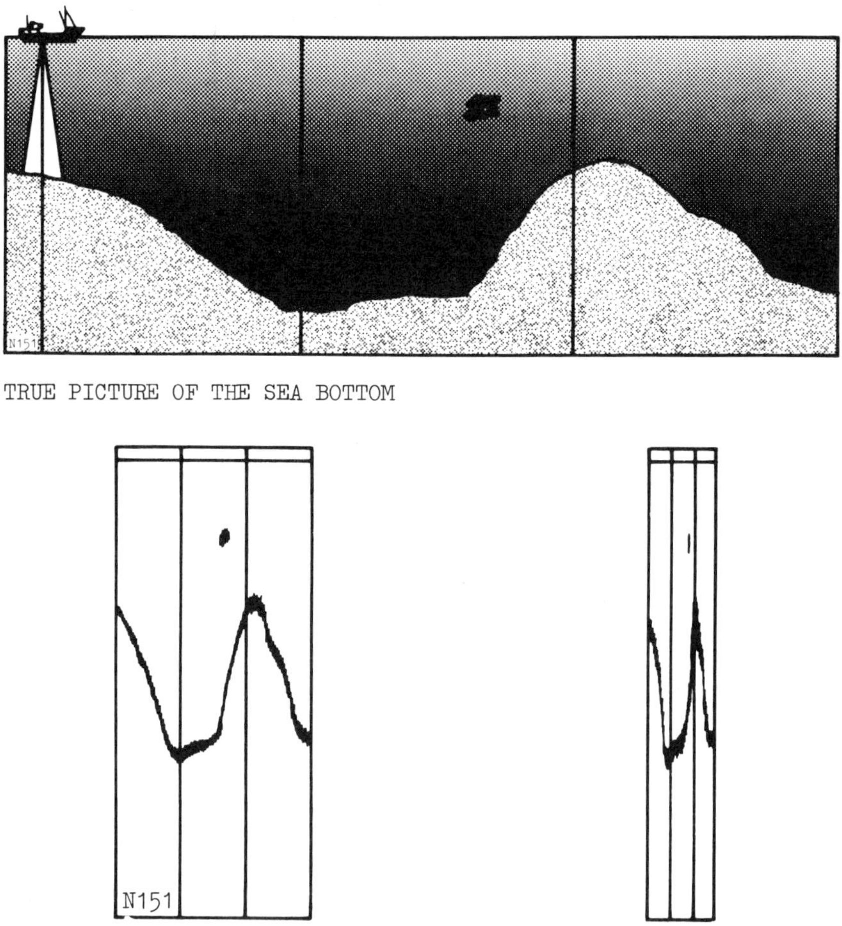

Fig 30 The relative dimensions of displays on an echogram using different paper speed or at different ship's speed. Left, low speed; right, higher speed

more readily compared. Narrow beam and short pulses with short 'rise time', will improve the ability to separate fish from bottom echoes. However, there are limits to both these improvements. Very short pulses can only be obtained with high frequencies and this limits the depth range, due to the correspondingly high rate of attenuation of the high frequencies in water. Very narrow beams require a stabilized transducer which involves a complicated and costly arrangement of control, either by mechanical methods or by a special multi-element transducer energized electronically on a correct time sequence. Simple gravity stabilization by use of gimbals has its limitations and is therefore seldom used.

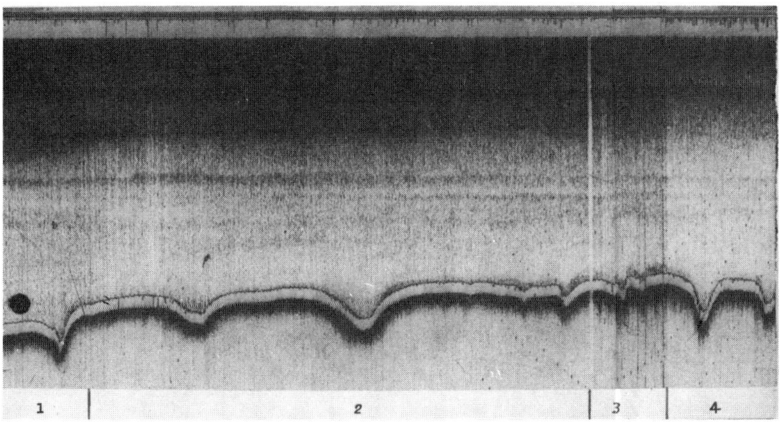

Fig 31 Exaggeration of sea-bed irregularities due to speed of vessel and paper speed

4.3 Scale expansion, bottom lock

In order to allow detailed characteristics of fish echoes to be studied, various methods have been developed whereby a smaller depth interval can be shown, either on the same paper record, on a separate recorder or on an oscilloscope (CRT). When bottom fishing, it is desirable to study the layer immediately above the bottom on a larger scale, but this is complicated by variations in water depth or by vertical movement of the vessel due to waves. To overcome this, more sophisticated echo sounders include a device which enables the display on a recorder or CRT to be locked to a preselected depth interval which can conveniently be locked on to the bottom, thus making the display independent of varying depth or ship's movements. This is called 'bottom lock' and is of considerable help in recognizing details of fish echoes near the bottom. *Fig 15* shows a bottom lock display on a CRT which indicates the characteristics of echoes from fish and bottom for each transmission. To avoid the need for continuous observation of a CRT, however, an echogram is required. As previously mentioned, some modern sounders combine the expanded-scale section and the normal depth display on the same chart. This combination, using a multi-stylus recorder, but without bottom lock, is shown in *Fig 18*. The alternative electronic method, needing only one stylus driven by a rotating belt, running at uniform speed, gives a similar but better display (*Fig 33*) because it avoids the white separation lines between the styli which are observable in *Fig 18*.

The electronic method uses a shift register as a memory. Echo signals are stored in digital form in the registers to be 'clocked' out later at a rate to suit the stylus speed and the desired expansion scale, which is considerably larger than the normal scale recorded on the upper section of *Fig 33*.

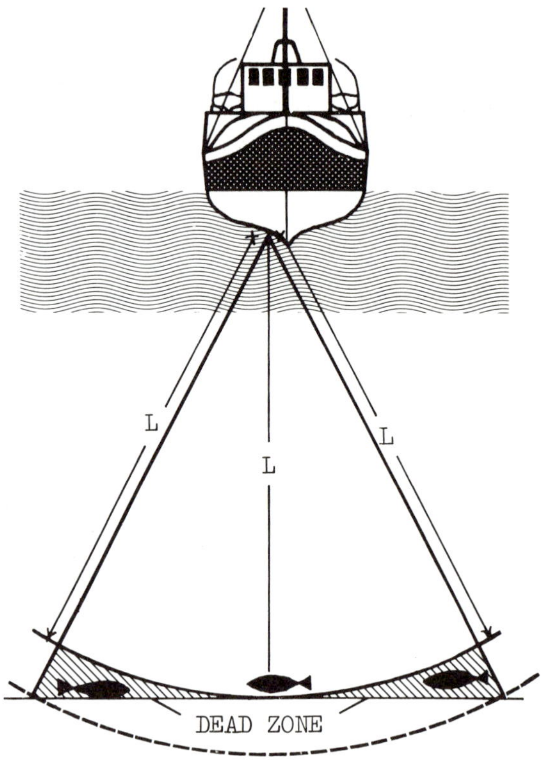

Fig 32 Diagram illustrating the 'dead field' in the sound beam, close to the bottom

4.4 Gain-depth relationship

Below the depth range of the initial distance-related gain reduction (about the first 30 to 50m) most echo sounders will amplify signals from any greater depth at the same rate. But because of beam spreading and attenuation, echoes from the same objects will decrease in strength with depth (Chapter 1). In order to give comparable recordings, the gain settings should be increased as the depth of the section being observed increases.

This can be seen from the sonar equation for the noise limited case which is:

$$SL - 2TL + TS = NL - DI + DT$$

SL = Source Level;
TL = Transmission Loss;
TS = Target Strength, *ie* its reflective property;
NL = Noise Level;
DI = Directivity Index of the receiving transducer;
DT = Detection Threshold, *ie* the required signal strength above noise to allow detection.

All units are in db (decibels) relative to 1 dyne/cm^2

In this equation, the received echo level is on the left hand side and, to be detected, this signal has to be equal or stronger than the noise being received and the detection capability of the equipment. Only the term TL is dependent upon the range and therefore, if the gain is varied according to 2TL, the echo will be amplified so that the recorded signal is independent

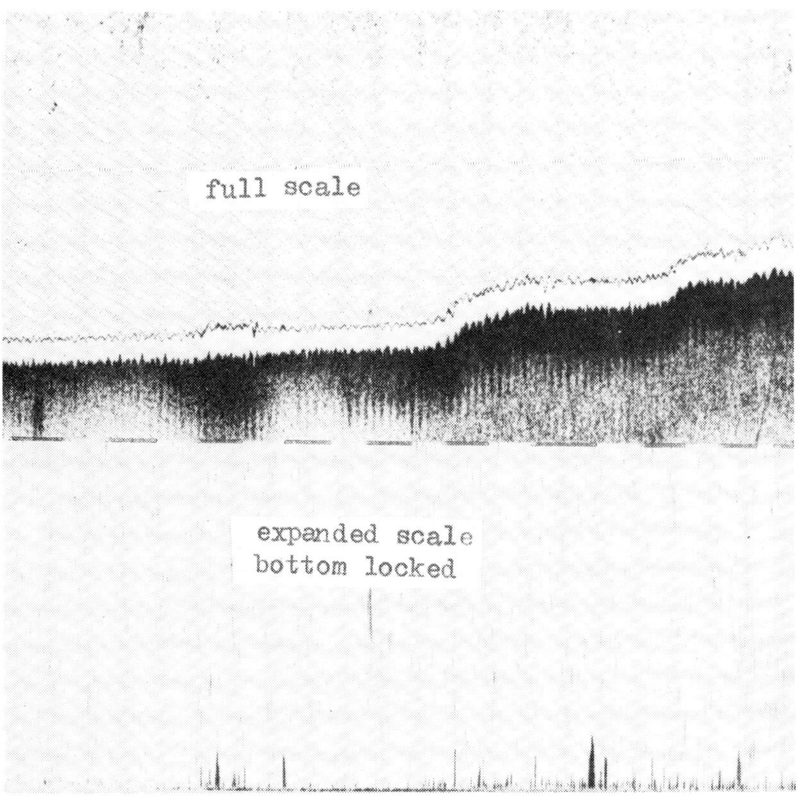

Fig 33 This echogram with bottom-locked scale expansion shows a feature of special interest, namely that where there were fish echoes close to the bottom on the expanded recording, and an intermediate degree of shading on the conventional recording, then the catch almost invariably turned out to be flatfish

of range. Unfortunately, the transmission loss for a small target within the transducer beam width (*eg* a single fish) is not the same as that for a target without boundaries within the beam width (*eg* fish school or sea bed). For single fish, the gain (G) would need to be:

$$G = K + 40 \log R + 2aR$$

and for a school of fish or the sea bed:

$$G = K + 20 \log R + 2aR$$

where aR is the attenuation in db/m, R is the range and K is a constant.

These equations are introduced here for those interested and are further explained in text books such as those by D G Tucker and B K Gazey or by R J Urick (see *Further Reading*). A complete understanding of these equations is not necessary for comprehension of this manual.

Varying the gain manually requires attention by the user and considerable experience for correct interpretation. Some echo sounders therefore have a built-in automatic gain compensation for the spreading and attenuation of the sound with depth. This is based on a system of time-varied gain (TVG). A TVG range of, for instance, 3 to 300m means that full amplification is applied to signals from 300m depth or greater, while signals from depths less than 300m will be amplified at reduced rates depending on their distance (depth). To have objects of the same target strength always displayed at the same level is of particular advantage when searching for fish over a substantial depth range. *Fig 34* shows an example of recordings made simultaneously on an ordinary echo sounder (a) and on one fitted with TVG (b).

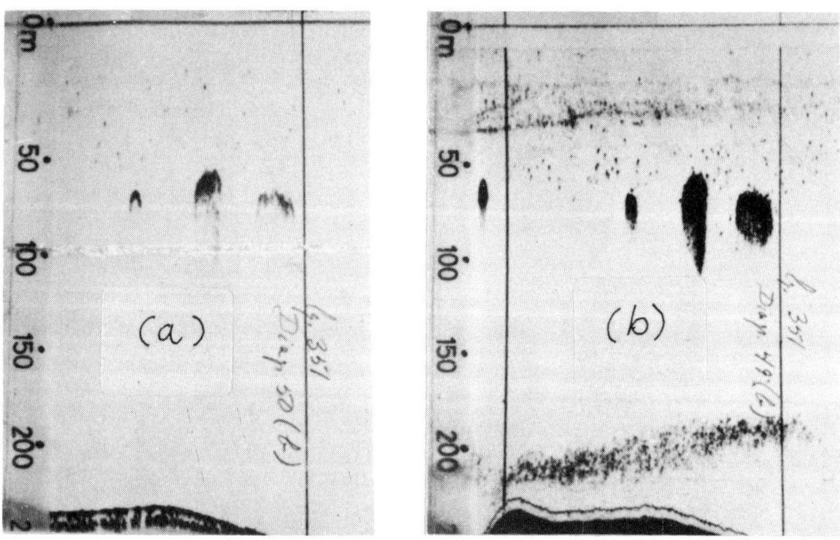

Fig 34 Comparison of echograms of the same targets, namely fish, scattering layer, and sea bed, (a) without and (b) with time-varied gain (TVG)

CHAPTER 5

DIFFERENT TYPES OF ECHO SOUNDERS

A number of different types of echo sounders are now produced for different purposes and requirements. The main factors which determine price, complexity and purpose of use are classified below:

Frequency	Low 10—20kHz Medium 20—100kHz High 100—400kHz
Multi-frequency	
Beam width	Wide 20°—30° Medium 10°—20° Narrow 4°—10° (narrower with stabilization)
Pulse length	Long more than 2ms Medium 1—2ms Short 0.1—1ms
Transmission power	High 5—10kW Medium 1—5kW Low less than 1kW
Amplifier	Simple initial suppression Time varied gain (TVG) Frequency band width selection
Transducer	Unstabilized Stabilized
Signal presentation	Lamp display (flasher) Chart recorder, wet or dry paper, without or with whiteline or greyline Chart recorder with additional scale expansion, bottom lock and phased display CRT or fishlupe Chart recorder combined with CRT TV type display, with or without colour

5.1 Simple depth sounders

Since the bottom is a good sound reflector, an echo sounder for recording only low to medium water depths can be fairly simple. The following combination of properties for such an instrument would be satisfactory and cheapest:

Frequency:	medium
Beam width:	medium
Transmission power:	low
Signal presentation:	lamp display ('flasher')
Amplifier:	simple initial suppression
Signal treatment:	no special

5.2 Demersal and pelagic fish-finding echo sounders

Considerable care should be taken when choosing a fish-finding echo sounder. First of all the particular uses for which the sounder is intended should be clearly appreciated.

A fish-finding echo sounder for surface layers and shallow depths (down to about 100m) should preferably have the following characteristics:

Frequency:	high
Beam width:	medium or narrow
Transmitting power:	medium or low
Pulse length:	short
Amplifier:	preferably with TVG
Signal presentation:	recorder with appropriate choice of ranges for pelagic and demersal fish and whiteline or greyline display

An echo sounder intended for use at medium depths (*ie* down to about 500m) should in general have the characteristics listed above for a shallow water or surface sounder, but with medium sound frequency, higher transmission power, bottom locked scale expansion and the ability to phase the short range display (*2.1.4*). Combination with CRT is desirable.

Fish detection at depths greater than about 500m requires high transmission power, a narrow beam and a medium to low frequency. Display should be the same as for medium range sounders, and stabilization of the beam is of great advantage. This means a costlier and more sophisticated instrument.

In general the cost of an echo sounder will increase with the depth at which fish have to be detected.

5.3 Bottom echo sounders for great depths

For the recording of the bottom at great depths, say several thousand metres, energy emission and sensitivity of reception must be as high as possible and attenuation losses minimal. Thus the frequency used must be low, the pulse long and the beam width narrow, necessitating a large, stabilized transducer. Such a system has rather poor resolution, and it is therefore difficult and expensive to combine a good deep sea sounder with the requirements of efficient fish detection.

5.4 Netsonde

For use in trawling, special acoustic instrumentation has been developed to measure the fishing depth and the opening of the trawl as well as recording the fish which enter the trawl. With more sophisticated units, the distribution of fish around and in front of the trawl can also be observed.

The netsonde was initially developed for midwater trawling. Later it was found that it can also be most useful in bottom trawling for measuring the opening height of the trawl mouth and for ensuring continuous bottom contact, *eg* when fishing over uneven grounds or along steep slopes. In addition to this continuous control of gear performance the netsonde used in bottom trawling also provides a much better estimate of catching efficiency and amount of catch by showing the fish in and above the net opening.

Netsonde equipment requires a direct connection between fishing gear and vessel which can be either by cable or by a wireless acoustic link with a transmitter on the gear directed towards the vessel and a receiving transducer mounted on the hull or towed in a paravane behind the vessel. For observing the trawl opening and fish entering, only one transducer is mounted on the headline of the trawl, usually sounding straight downwards. When midwater trawling over very deep water with the bottom outside the sounding range the transducer may be attached to the groundrope sounding straight up to the surface.

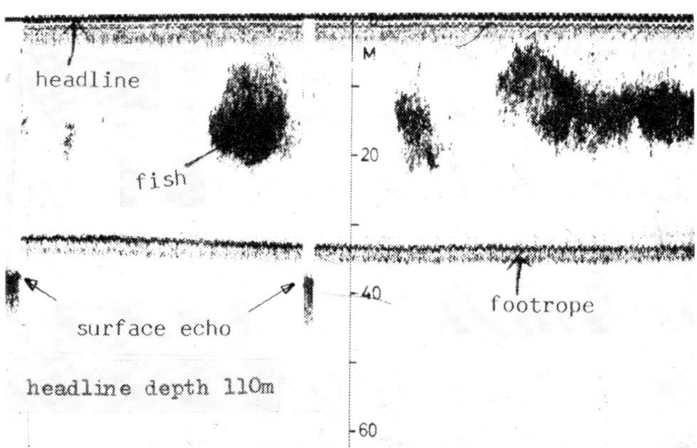

Fig 35 Netsonde echogram from headline-mounted transducers. The two narrow bands (left) show the depth to the headline (110m). The opening of the trawl (33m) and the fish being caught are also shown. The bottom is outside the recording range. Note: the recording ranges for upward and downward sounding are not the same

The wireless acoustic link can be successfully used in medium depth and under good alignment and low noise conditions. For greater trawl depths or unfavourable conditions, as well as for 'multi-netsonde' systems, a cable connection between the trawl and the trawler is preferable, being much simpler, less subject to interference and easier to maintain in an operational condition. For limited depths down to about 100m, using a relatively short cable (300 to 500m), any medium size echo sounder can be made into a combination ship's sounder-cum-simple netsonde. All that is needed is an additional transducer for the net, a suitable cable (2 core), a simple winch and a switch to shift at will from the ship's to the net transducer. *Fig 35* shows an echogram obtained from two transducers mounted on the headline of a midwater trawl and used alternately for upward and downward soundings. *Fig 36* is an example of an echogram obtained from a unit with an acoustic link, during bottom trawling.

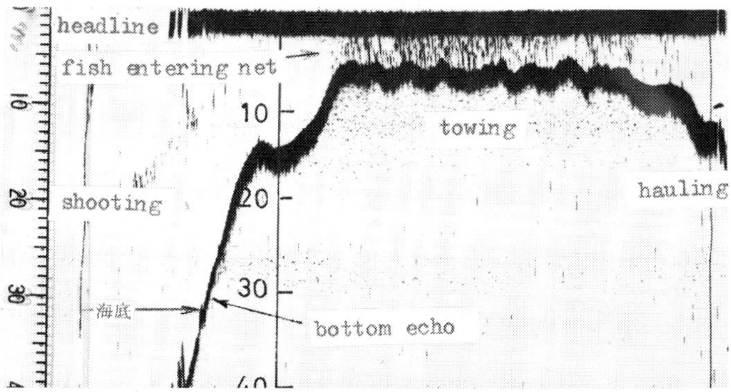

Fig 36 Example of echogram using an acoustic link netsonde, with a single headline transducer directed downwards, during bottom trawling. Since the headline transducer is mounted at the centre of the bosom, and the net has a long square, it is too far ahead of the groundrope to record it, since the latter is outside the beam

When operating a pelagic or midwater trawl it is essential that the skipper knows the position of the trawl relative to the fish, the bottom and the surface, so that he can 'aim' the trawl at the fish and so maximize the catch. Consequently, a further development of the netsonde attempts also to obtain information about the fish distribution and their reaction in front of and around the trawl, as well as providing more complete data on the gear itself. The method used is to mount a number of transducers at different parts of the trawl to record not only the trawl depth, trawl opening and fish below the headline, but also the distance between the trawl wings, and the distance and amount of fish recorded in various directions relative to the gear and inside the net (catch estimates). One obstacle to the effective use of such arrangements has been the interpretation of the information obtained when the recordings from the various transducers are only presented separately and intermittently. These difficulties can be overcome by using several display units, as are available in various research systems and also in some commercial multi-netsonde equipment. An example of the transducer arrangement for experimental multi-netsonde sounding is given in *Figs 37a* and *37b*.

Here the netsonde unit includes a switching arrangement by which the functioning of the various transducers is controlled at will. In one type of multi-netsonde equipment the signals from the transducers which are mounted vertically are displayed on an echo sounder recorder, while signals

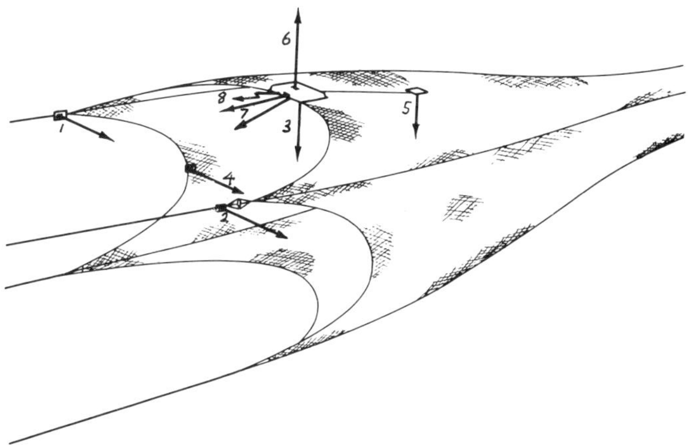

Fig 37a Schematic view of an experimental multi-netsonde transducer arrangement

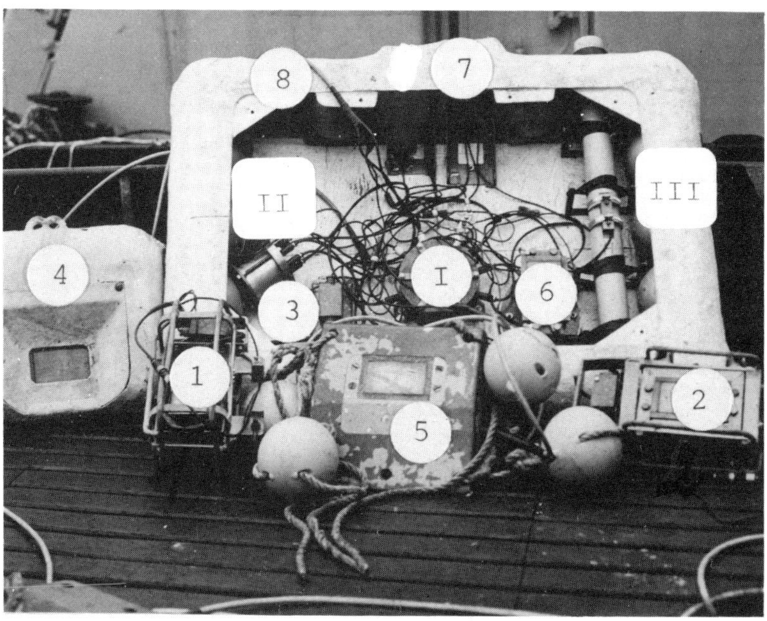

Fig 37b Multi-netsonde board (opened) for the trawl arrangement in *Fig 37a*, together with transducers, the numbering of which corresponds to those in *Fig 37a*. Other components are: I, main relays; II, relays for forward transducers and III, current meter, used as a log

from the transducers pointing in various directions in front of the trawl are shown on a CRT display. Using this display the skipper can observe whether a fish school being approached is higher or lower than the trawl opening or to the right or left, and an adjustment of trawl depth and course may be made accordingly.

Most commercial netsonde equipment has only one headline transducer sounding straight down. By removing the acoustic shielding on the backside of the transducer simultaneous sounding down and up can be achieved. This is the cheapest way of providing net depth control when midwater trawling over great depths and when the sea bed is outside the sounding range. This makeshift up-and-down netsonde has the disadvantage that fish traces which are shown below the headline may actually come from fish above the headline. To avoid this uncertainty two transducers are used in more sophisticated commercial sets. Still more sophisticated (and expensive) sets have additional forward transducers and a thermometer measuring the water temperature at the net for use when seeking fish with a narrow temperature preference. Net speed (through water) meters are not yet commercially available (*Figs 37a* and *b*). Netsonde equipment which is essential for 'aimed' midwater trawling, is used commercially with up to, and sometimes more than 2,000m cable length for trawling down to 500m and more.

CHAPTER 6

ECHOGRAMS AND THEIR INTERPRETATION

Normal echo sounding equipment is used in fishing to determine the depth of water, the nature of the sea bed and the position, quantity, species and sizes of fish present. The analysis of the echograms and their interpretation is not always easy and straightforward, because the nature of echograms is determined not only by the types of targets, but also by the characteristics of the respective acoustic instruments and by the operating conditions. The amount of information that can be obtained from echograms depends essentially on the skill and experience of the operator. In this chapter examples are given of the main characteristic types of traces.

6.1 Recording of soft and hard bottom; types of bottom and topography

Besides depth, which is easy to read, the echo sounder can give information on a number of characteristics of the sea bed itself (often called ground discrimination). The most important of these characteristics are, whether it is smooth or rough, soft or hard and how it slopes. *Fig 38* demonstrates several of these.

On the left of the echogram the vessel passed over level, smooth ground, which suddenly became rough and uneven. It can also be deduced from the trace that the smooth, even ground is soft and the uneven ground is hard, probably rocky. The difference in echo length or tails of the bottom trace indicate this. The soft ground echo was weaker, and therefore produced a narrower trace than the stronger echo reflected from the hard ground. The echo length of the bottom trace will also vary of course, with the depth, with the slope of the bottom and with the sensitivity (gain) used, but with some experience and training the operator will be able to judge the character of the sea bed from the echogram.

A further example of this type of echogram is shown in *Fig 39*, which also illustrates the value of the whiteline feature in showing fish close to the bottom as well as indicating the sea bed contour. The displayed depth variations of the sea bed need careful interpretation since they are related to the speed of the paper relative to the speed of the ship (*Figs 30, 31*).

One manufacturer has produced a transducer specially designed to improve ground discrimination (*Fig 40*). By mounting two transducers side by side in a common housing, they can be arranged so that their beam patterns interfere, effectively reducing the power in the main beam and increasing that in the side lobes. The angle at which the side lobes strike the

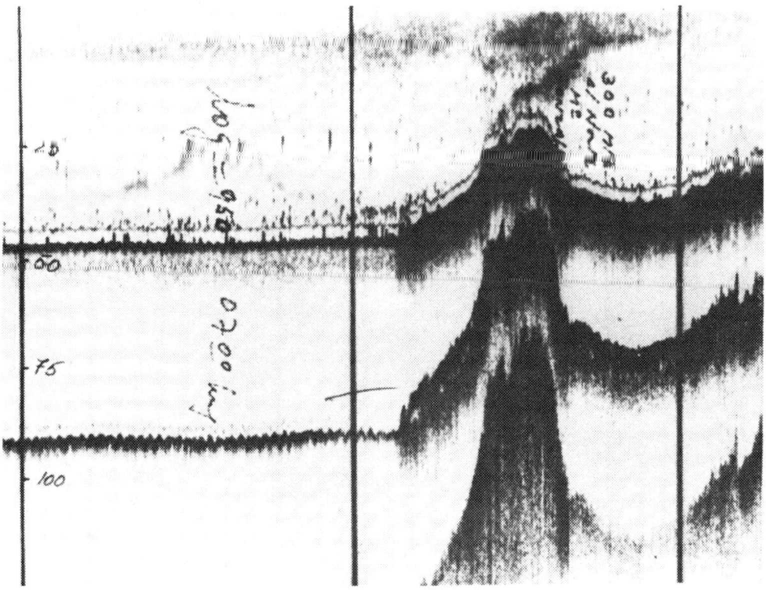

Fig 38 Echogram showing smooth, soft bottom on the left and uneven, hard bottom on the right. Note increased echo length and stronger double and triple echo from hard bottom

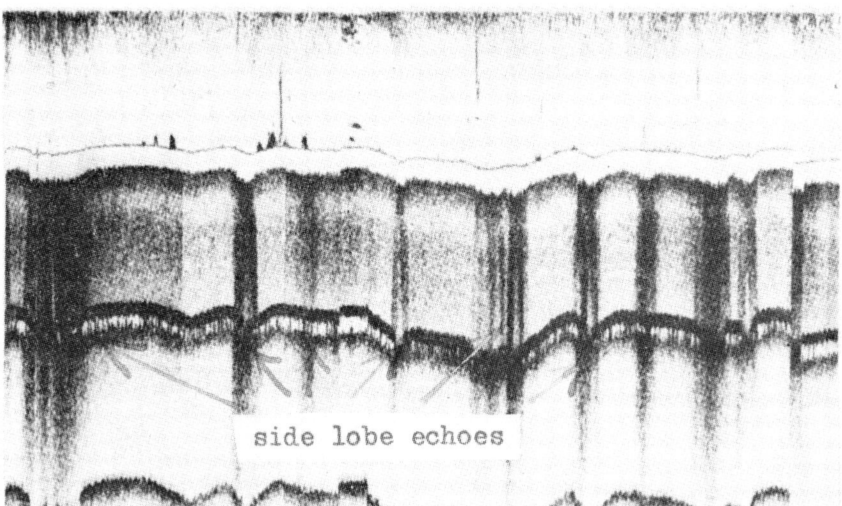

Fig 39 Echogram of hard and soft bottom in about 20fm depth, obtained by using a special transducer which improves ground discrimination

sea bed is such that, if the ground is smooth, most of the energy is reflected away, but if it is rough, more will be returned and displayed as the wider trace, *ie* a long tail following the bottom trace.

In addition to a long tail, the hardness of the ground is characterized by second and possibly further multiple echoes. These occur when the sound pulse which has been reflected from the sea bed back to the vessel (first echo), is then reflected down again by the surface and then returns, producing a second bottom trace indicated at roughly double the depth. A whole series of such multiple bottom traces may be recorded, depending on sound output, gain, distance and target strength. This effect is therefore particularly common when transmitting high power in shallow water on hard bottom. Very dense fish schools may also produce a second (double) trace, but in most cases a trace on the bottom looking like fish can be identified as rock or obstacle by checking whether it gives multiple echoes or not.

Multiple bottom traces can cause confusion when using a phased display or when the depth is beyond the depth scale set on the recorder. The echo from the true bottom may be recorded at an incorrect scale or a double echo may be misinterpreted as the true first bottom echo. When the bottom is very much beyond the set scale the pen may even return to the zero position before the first bottom echo returns so that this is recorded after a new transmission is made. *Fig 41* shows an example of such a 'ghost bottom'. This one is easily recognizable because the trace passes the zero

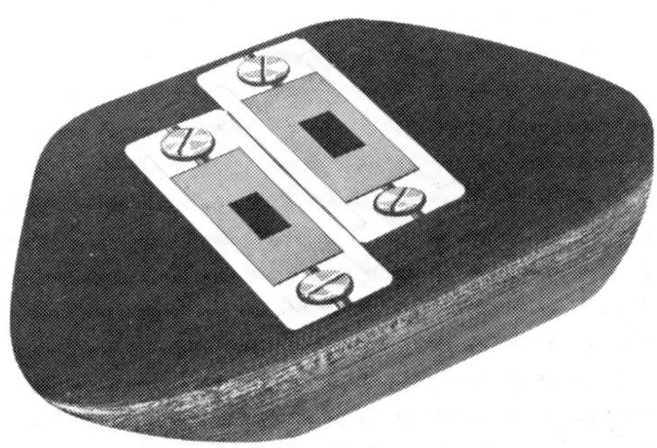

Fig 40 A 48kHz 'Z' transducer system in a wooden fairing

line and because it disappears at very shallow depths. This is caused by the initial suppression of the amplification (*2.1.3*) and demonstrates that the echoes were weak and therefore must have come from far away.

Another indication of the weaker echo from a 'ghost bottom' as compared to that from a true, shallow bottom is that the whiteline effect will usually not occur in the 'ghost bottom' trace.

'Ghost bottoms' from first or multiple bottom echo traces have given rise to many reports of new, hitherto unknown, shallow banks and great fish concentrations which have mysteriously disappeared on further investigation. An experienced operator can easily recognize 'ghost bottom' traces by the above signs, and by switching to deeper or suitably phased ranges as available.

Greater sub-bottom detail can be obtained by using a lower sound frequency. There are special echo sounders with high frequency for recording the surface of even very soft mud and low frequency for deep bottom penetration. Also, electronic techniques are applied to this effect as shown in *Fig 42*.

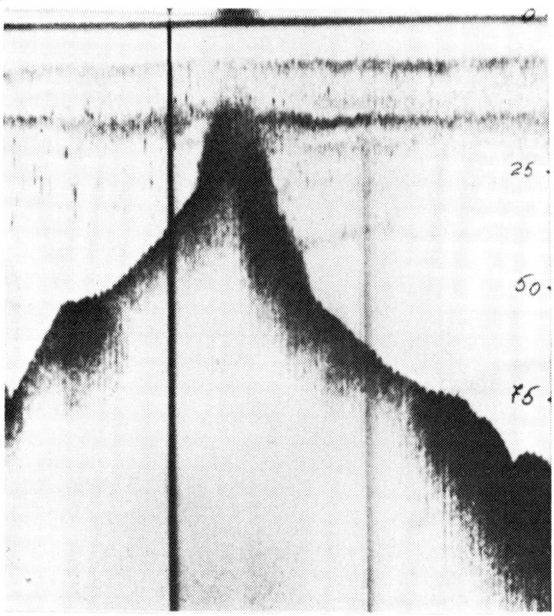

Fig 41 Echogram of 'ghost bottom'

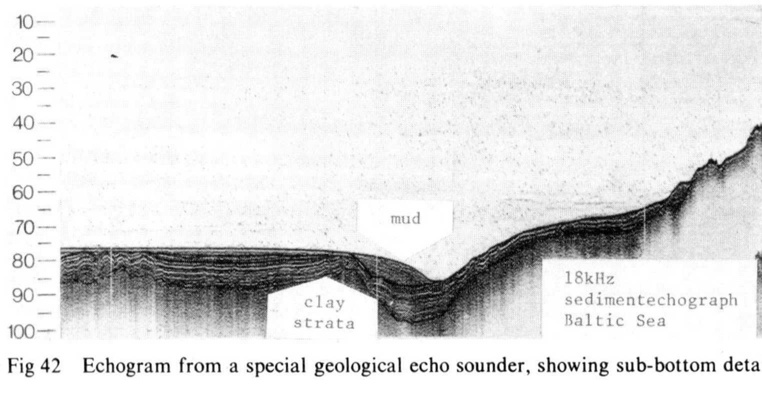

Fig 42 Echogram from a special geological echo sounder, showing sub-bottom detail

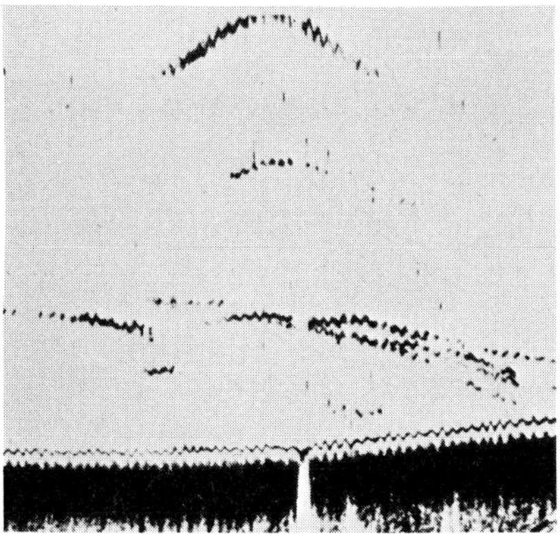

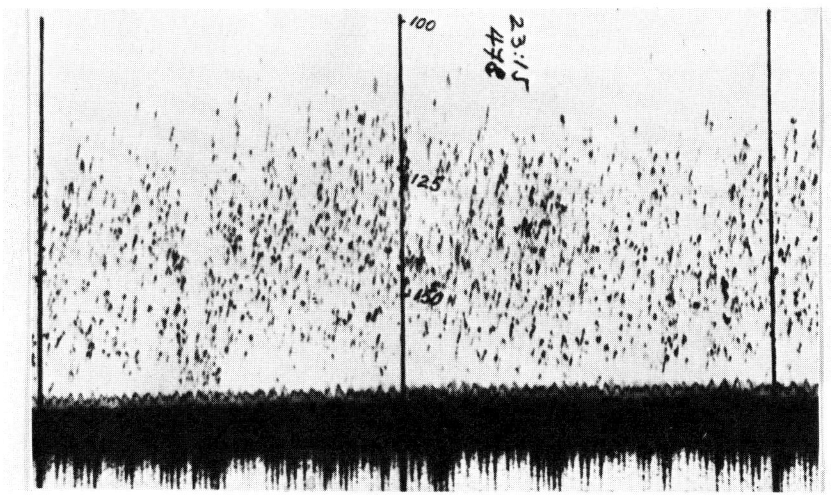

Fig 43a Single fish traces obtained when both vessel and fish were stationary. Fig 43b Single fish traces made with vessel at slow speed. Note 'inverted V'. Fig 43c Single fish traces made with vessel at high speed

6.2 Echoes from single fish

The traces of single fish may vary in appearance according to types of echo sounder and varying operating conditions. With the vessel lying stationary, single fish remaining in the sound beam will be recorded as continuous line traces, as shown in *Fig 43a*. At low speeds, the lines shorten and become curved (*Fig 43b*) and at high speeds they appear as dots or points (*Fig 43c*). The reason for this is that the number of echoes obtained from each individual fish decreases as its relative speed of travel through the sound beam increases. The beam width in the ship's direction of travel will also

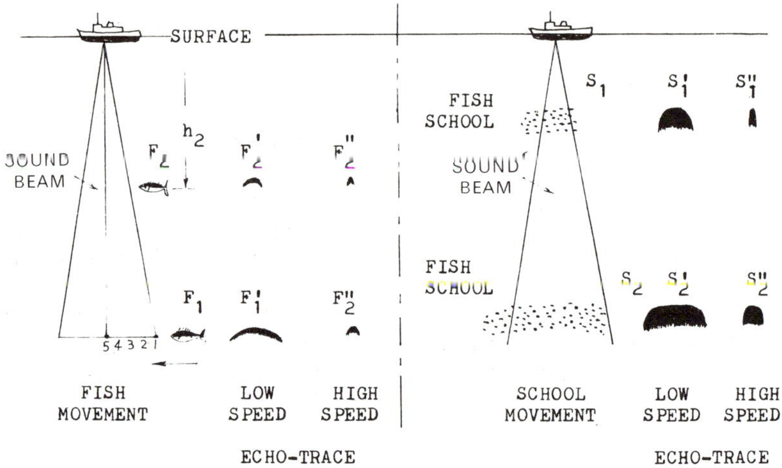

Fig 44 Formation of traces of individual fish and of fish schools

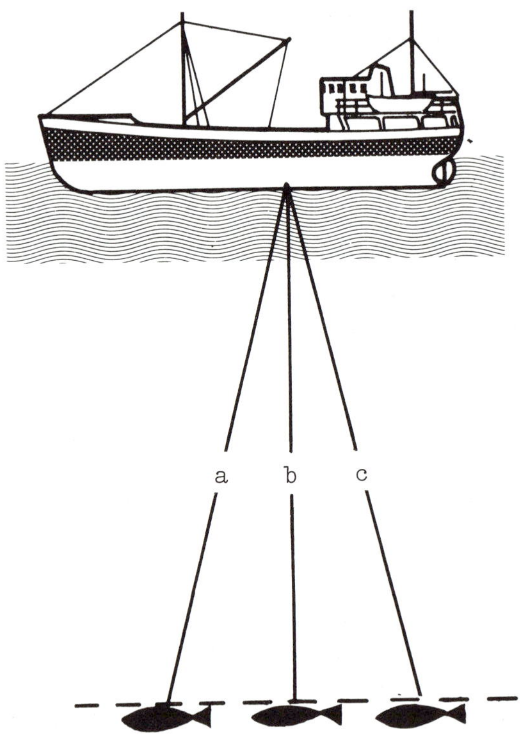

Fig 45 One fish in three different positions during passage of sound beam. The oblique distances *a* and *c* are greater than the vertical distance *b*

influence the appearance of the single fish trace. The wider the beam, the longer will be the individual traces (*Fig 44*).

The curved line or inverted V-shaped traces appearing in recordings at the relatively low towing speed (or with a wide sound beam, even at higher speed) are typical of single fish traces. They are caused by the fact that the distance between the fish and the transducer is greater when the fish is near the edge of the beam than when it is directly under the ship, as seen in *Figs 44* and *45*. When the ship is pitching or rolling this characteristic shape may be difficult to detect. In order to identify the kind of fish being recorded it is usually necessary that a well defined trace of individual fish is produced, *ie* a succession of echoes from the same individual which take the form of a compact continuous trace. The beam width should therefore be related to the sounding rate and the paper speed in such a way that several echoes are obtained from individual fish at normal speeds.

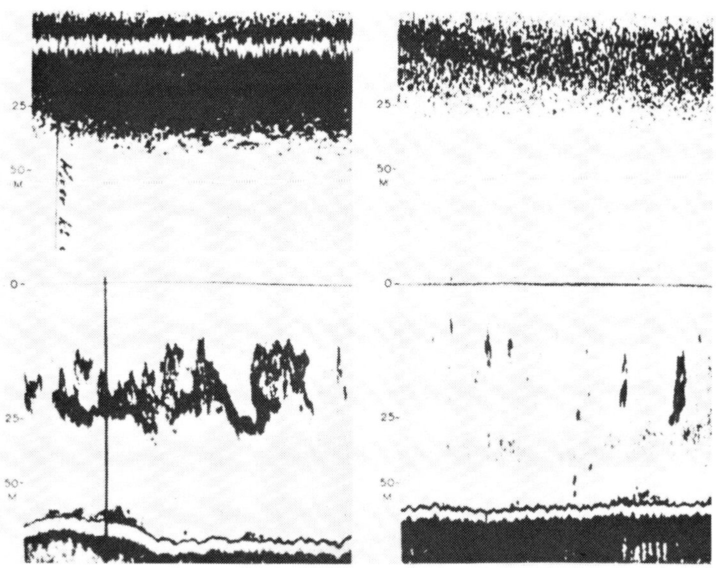

Fig 46a Echograms of different types of layers (top) and below, schools of small herring

6.3 Recording of fish in layers and schools

When the fish density exceeds the limit of the resolution of the echo sounder, the records overlap to form a continuing trace, which may appear as a layer or as a distinct solid blot. There is, of course, a continuous transition from dispersed layers or bands, where the echoes from a large number of single individuals can still be identified, to dense layers and solid school traces. *Fig 46a* shows some examples of different types of layers and distinct schools of small herring, while the lower section of *Fig 46b* shows the same for adult herring.

A common feature of the layer type traces which are often produced by

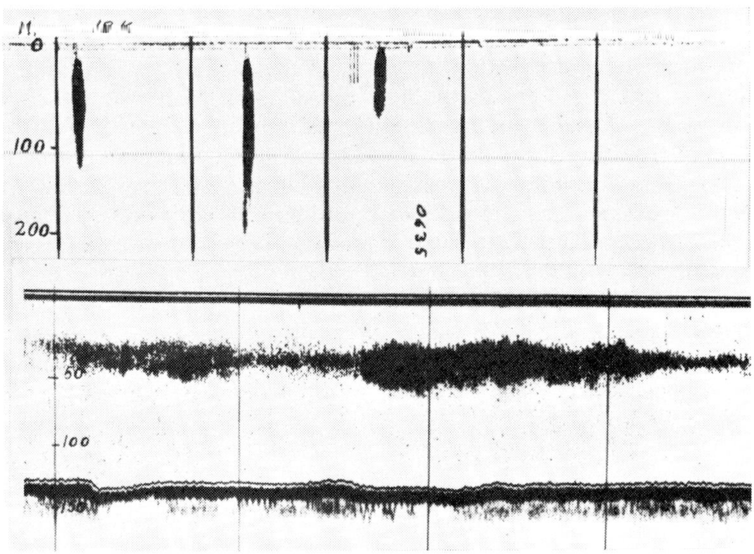

Fig 46b Echograms of schools (above) and layers (below) of large, adult herring

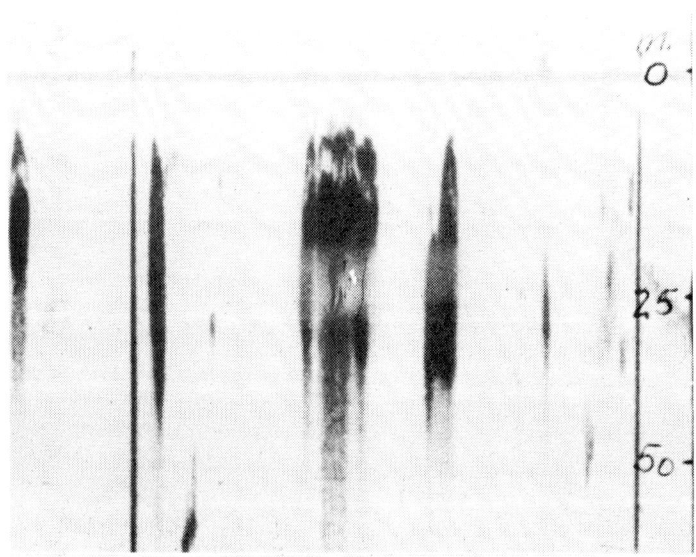

Fig 47 Echogram of dense schools of anchovy

fish is the presence of single fish traces at the upper and lower borders of the layer. This is clearly shown in the upper section of *Fig 46a*, and in the lower section of *Fig 46b*.

Well-defined schools of pelagic fish do not usually 'thin-out' at the edges as with the layer traces. This is shown by the typical traces of herring schools in the upper section of *Fig 46b*, and again in *Fig 47*, which shows schools of anchovy. A typical feature of echograms of dense pelagic schools is their echo length or 'tail' caused by sound reverberation. The sound-pulse is partly scattered inside the school and some of it is reflected back to the transducer even after the return of the direct echo of the deepest part of the school. While the upper edge of fish schools is shown in the correct depth, the lower edge is always shown too deep and therefore the vertical length of the trace of a school is not a direct measure of the true vertical extent of that school.

Further characteristic fish echograms are shown in *Figs 48* to *61* with explanations and comments given in the legend.

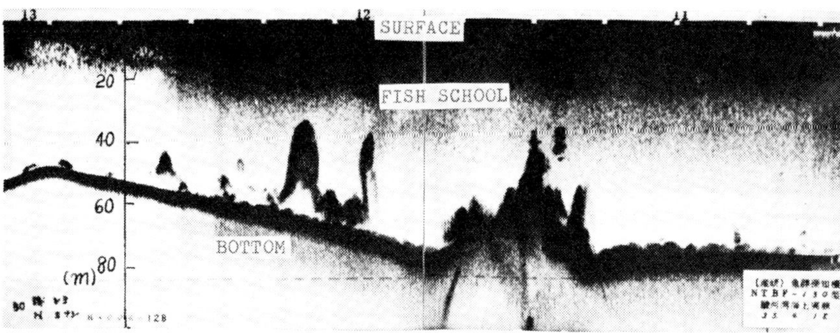

Fig 48 Echogram of schools of fish on and off the bottom. Without whiteline or greyline, bottom fish are not easy to recognize

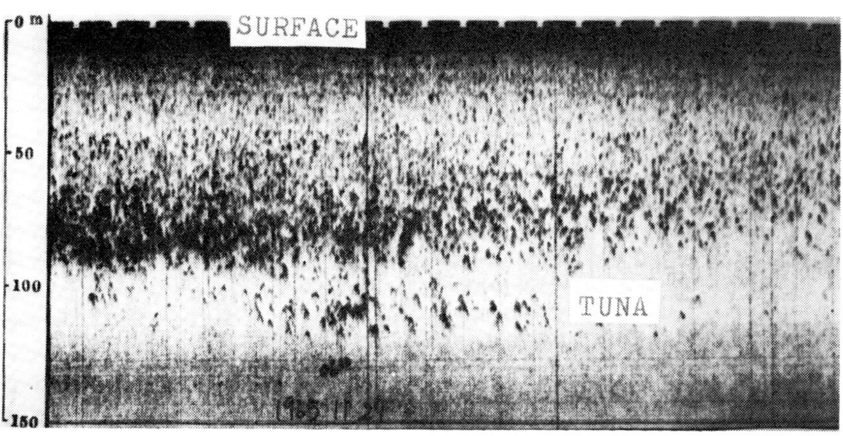

Fig 49 Echogram of tuna, from the mid Pacific Ocean

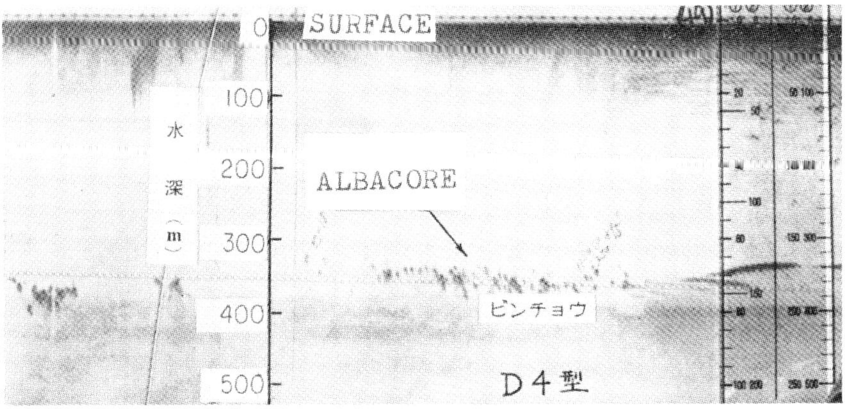

Fig 50 Echogram of albacore tuna at a depth of 350m

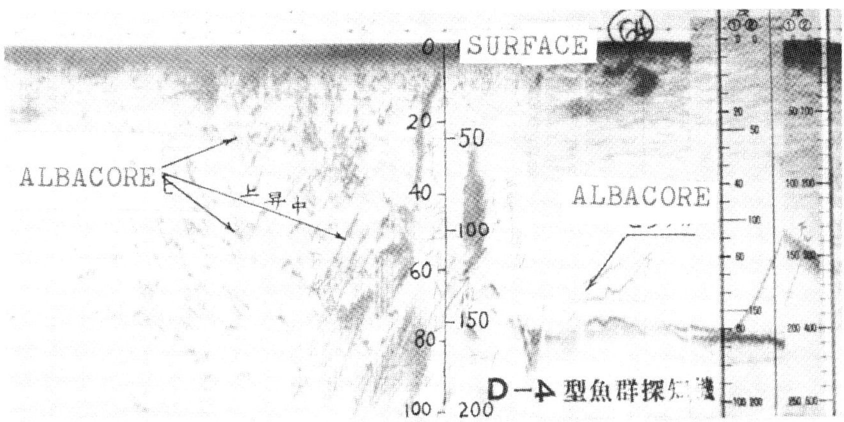

Fig 51 Echogram of albacore tuna rising rapidly to the water spray and bait fish of a pole-and-line fishing vessel

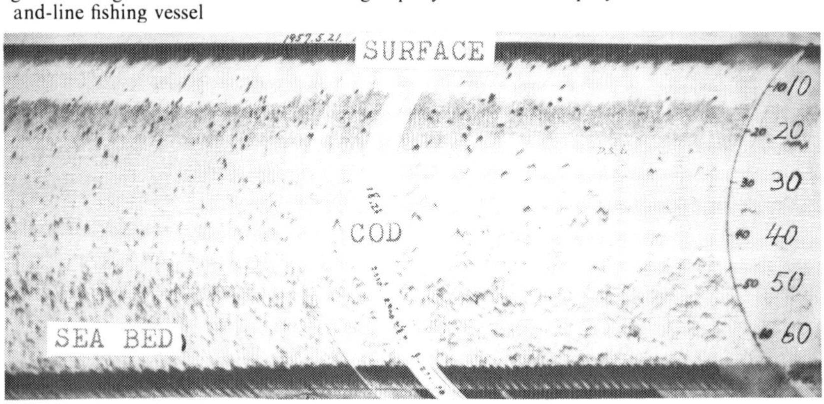

Fig 52 Echogram of single cod made by a 50kHz echo sounder in Okhotsk Sea. Note 'windmill' recording

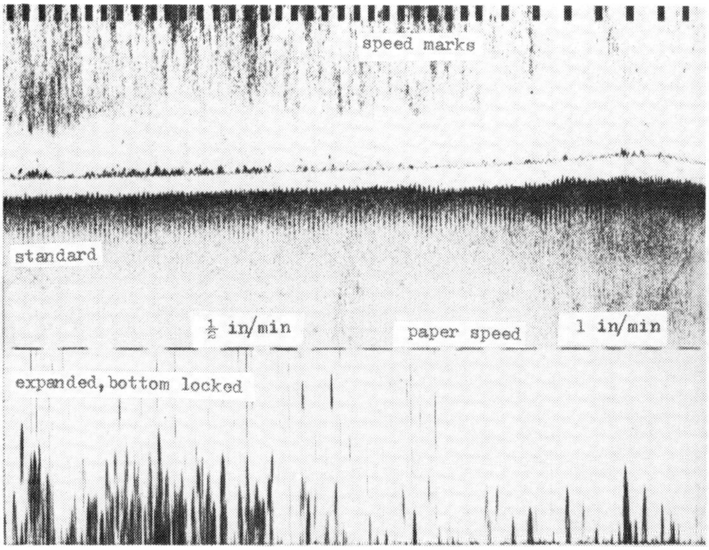

Fig 53 Echogram with bottom-locked scale expansion (below) from Iceland. Depth, 45fm; scale of standard record, 90fm across full paper width; scale of expanded recording, 2½fm across 3in; paper speed, ½in and 1in/min; ship speed, 4 knots; fish, cod

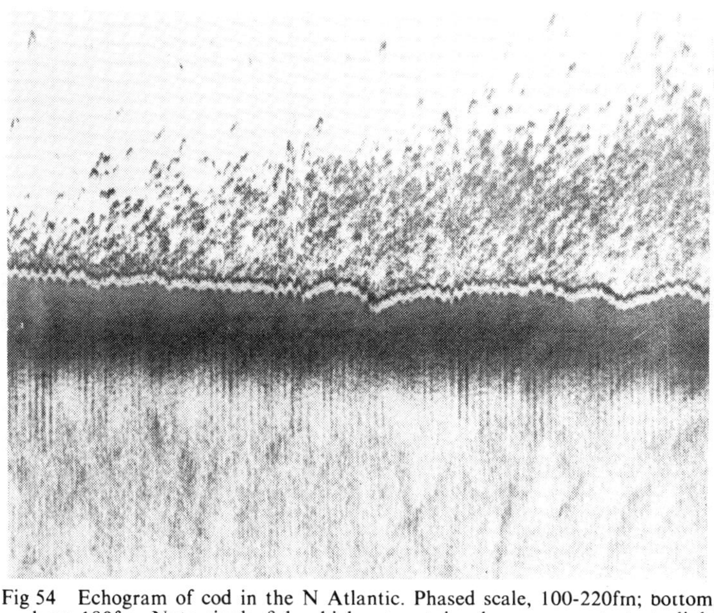

Fig 54 Echogram of cod in the N Atlantic. Phased scale, 100-220fm; bottom, about 180fm. Note single fish which, at certain places, concentrate to light schools

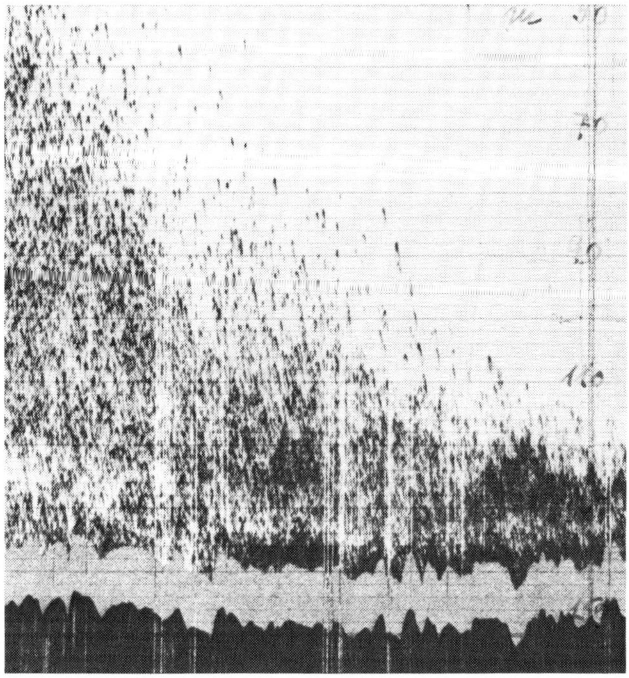

Fig 55 Echogram showing small redfish in the N Atlantic. Phased scale, 50—150m; ship's speed, 3.5 knots

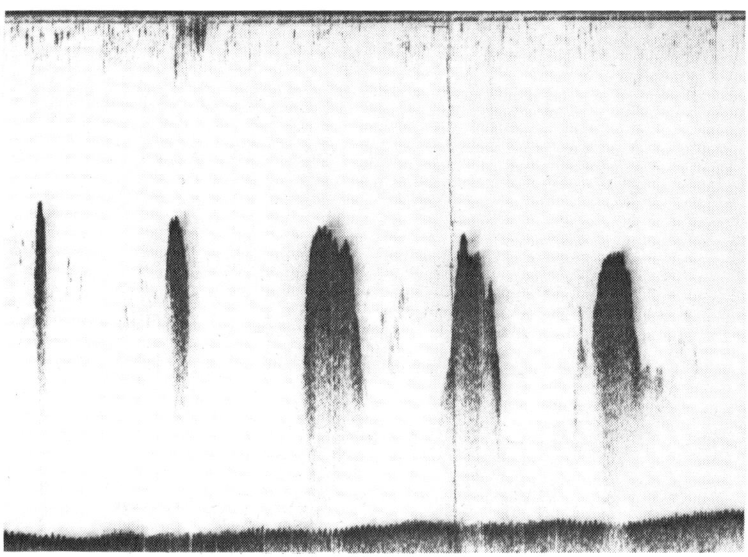

Fig 56 Echogram showing sprat schools in the North Sea. Scale across chart, 60fm; depth, 50fm; paper speed, 0.32in/min; ship's speed, varying 8—3 knots

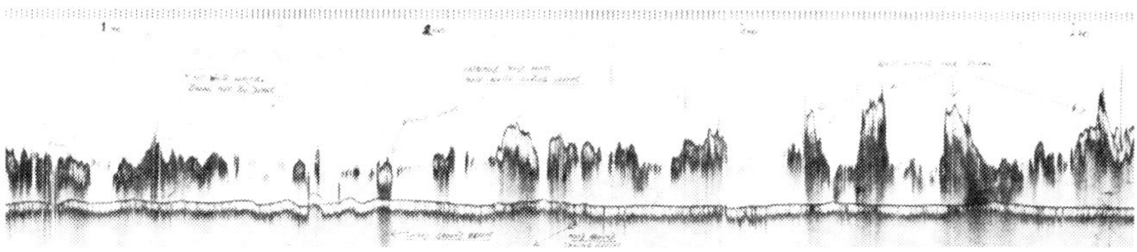

Fig 57 Echogram of sprat schools in the North Sea. Depth of water, 60fm; scale across paper, 90fm; paper speed, ½in/min; ship's speed, 4½ knots. Note whiteline recording of particularly dense parts of many of the schools

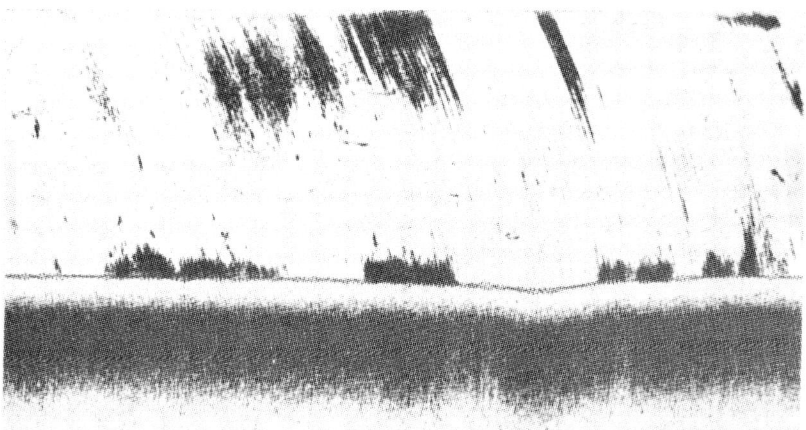

Fig 58 Echogram of dense prawn schools in the Gulf of Carpentaria, N Australia. The echoes close to the surface are propeller wakes from other vessels. Note that this echo sounder does not record with straight, but with curved lines, indicating that the stylus is not driven by a belt, but by a rotating disc or lever. This is 'windmill' recording, the curvature of which is in the opposite direction, to that in *Fig 52*

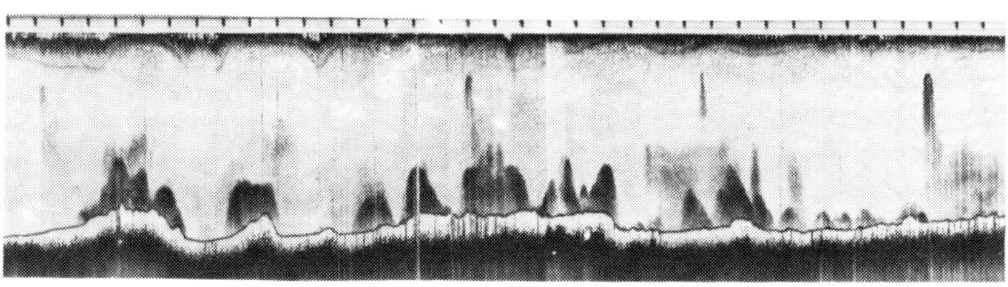

Fig 59 Echogram of sea bream schools off Niigata, Japan. Range, 0-160m

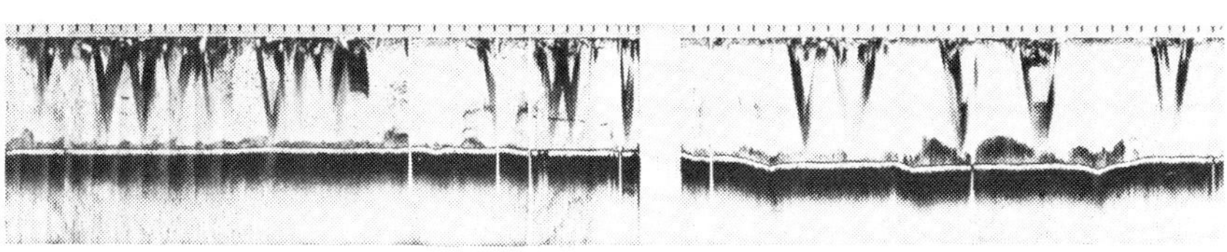

Fig 60 Echograms of squid off Toyama Bay, Japan. The strong V-shaped traces are caused by the jigging lines. Frequency, 50kHz; range, 0—200m

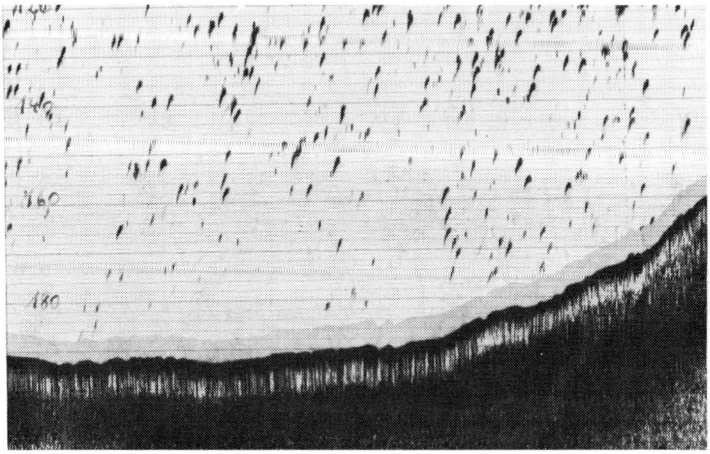

Fig 61 Echogram showing individual cod off the Norwegian coast. Note greyline recording of bottom trace as compared to whiteline in other echograms. The long ascent and short descent of the single fish traces is due to the slightly forward-pointing, oblique installation of the transducer *(3.1 Fig 21)*

6.4 Recording of plankton and other small organisms

In addition to the layer-type traces caused by fish schools as described above, other types of diffuse layer traces are also commonly produced by dense concentrations of smaller organisms, for example, small fish species such as lantern fish, larvae and plankton. They are often associated with a

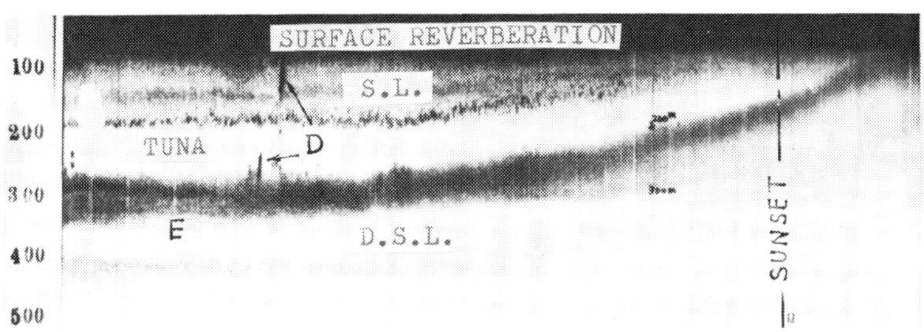

Fig 62 Echogram showing scattering layer (SL) and deep scattering layer (DSL) ascending at dusk. 'D' indicates tuna echoes

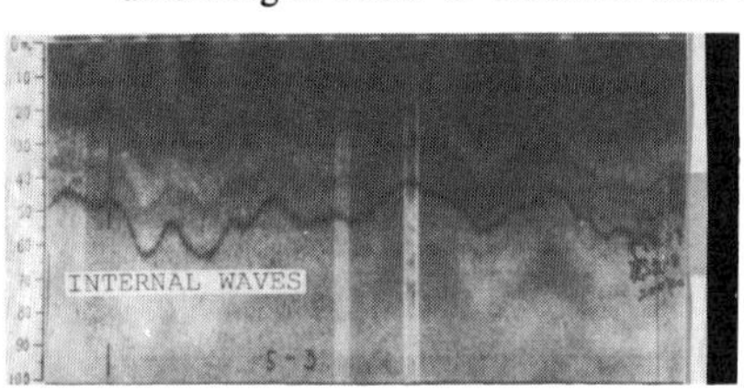

Fig 63 Echogram showing dense internal wave in midwater

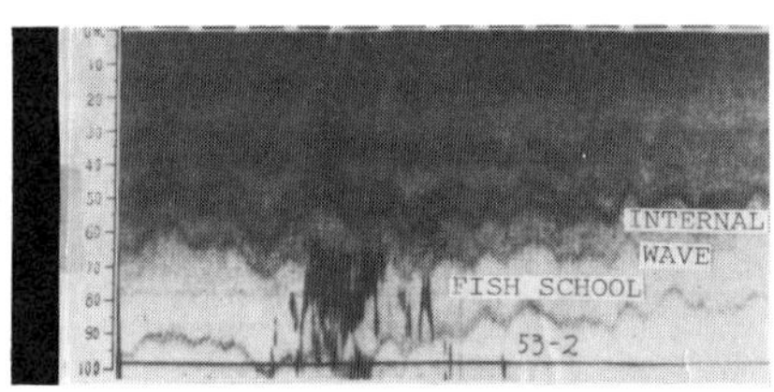

Fig 64 Echogram of fish school associated with internal waves

discontinuity layer (temperature, salinity) in the water column and appear as very finely grained 'clouds' or 'layers' which may cover large areas. Such layer traces are produced by organisms whose individual reflecting ability is close to or below the detectable limit of the echo sounder, but when occurring at high densities give the characteristic diffuse traces called scattering layers (SL) or deep scattering layers (DSL).

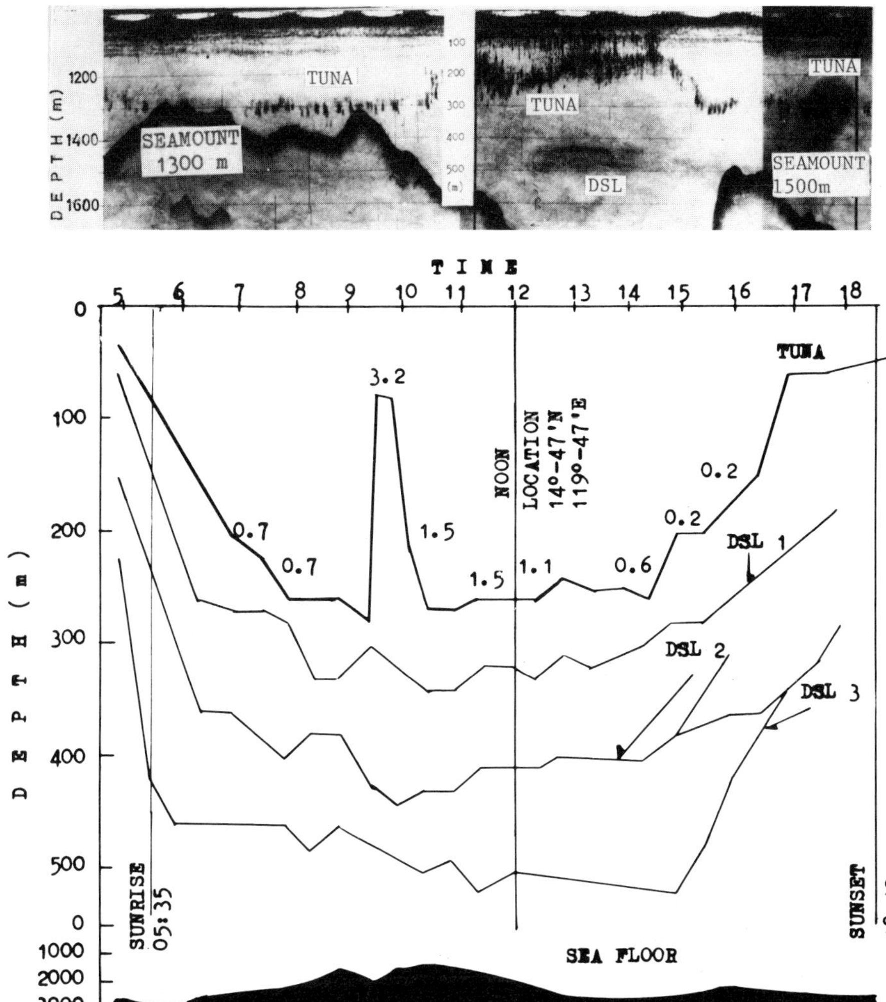

Fig 65 Above: Echogram of tuna distribution around a sea mount. Note recording in two depth layers with 1,000m difference *(6.1)*. Frequency, 14kHz. Below: diagram of diurnal depth changes of tuna (thick line) and of three deep scattering layers (DSL, thin lines) observed with 14kHz echo sounder in the South China Sea

They can often be distinguished from the layer type traces produced by larger fish because there are usually no traces of individual fish in association with the main layer traces. Furthermore, the character of the trace does not change significantly if the vessel slows down or stops.

The echo from scattering layers is stronger when higher frequency sounders are used, because the shorter wavelength is more suitable for small targets. An example of an echogram illustrating scattering layers and their depth variations with daylight is given in *Fig 62*.

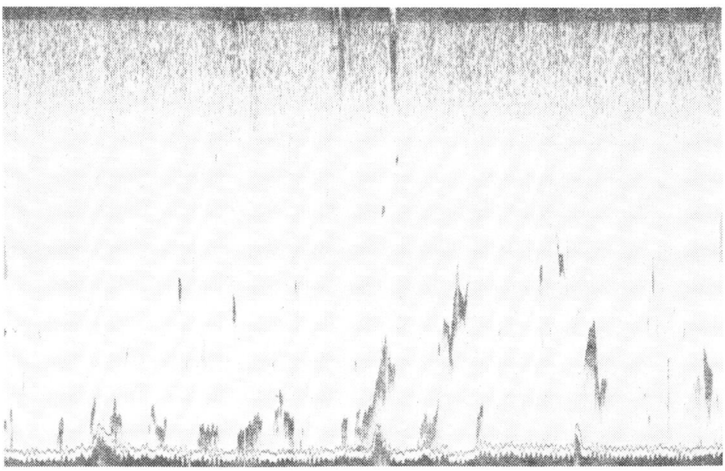

Fig 66 Scattered horse mackerel traces on the bottom during day-time. Traces in midwater are probably pilchard

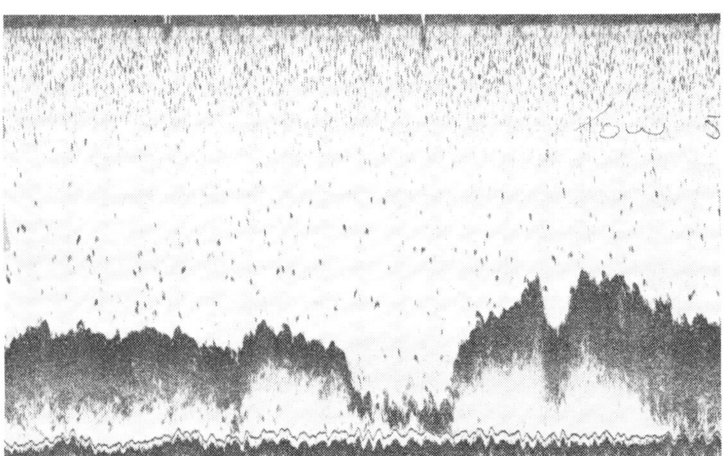

Fig 67 Typical night 'layer' trace of horse mackerel

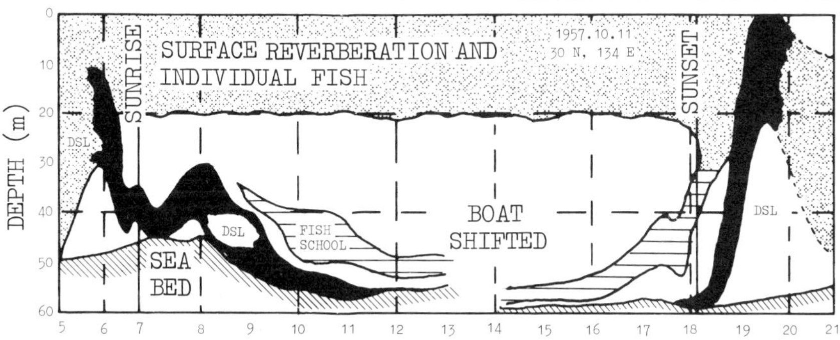

Fig 68 Diagram of diurnal changes of depth and distribution of fish in relation to DSL observed by echo sounding in the East China Sea

6.5 Internal waves

The echogram sometimes indicates an irregular periodic trace as in *Fig 63*. This may be caused by the boundaries between great differences in water densities, which are related to temperature and salinity, being moved by small variations of current. The latter effect is referred to as 'internal waves'. Where such wave-like traces appear, there is often a fish school either below or between two such layers. *Fig 64* is an example from the western part of the Indian Ocean, showing such a fish concentration. It has been observed that the catch of tuna was low when the internal wave effect appeared.

6.6 Fish distribution and behaviour

Echo sounding is of great value for studying fish behaviour such as changes in distribution occurring throughout a 24 hour period (diurnal migrations). *Fig 65* shows the diurnal change of the swimming depth of tuna, which reached its lowest point at midday. At 10.00am the tuna came nearer to the surface and this was at a point over a sea mount. Japanese longline fishermen seek out the shallower water over sea mounts to obtain good tuna catches. *Figs 66* and *67* show a similar day and night effect for horse mackerel.

Fig 68 shows a more complex diurnal effect for a fish school in relation to a DSL observed in the East China Sea. At night only individual fish scattered in the DSL were observed, but at dawn, the DSL began to descend and the fish started to school and pursue it. The converse effect occurred at dusk, and during the day, the fish school remained at a fairly constant depth close to the sea bed. This illustrates how it is possible to obtain good catches during the day with a bottom trawl and at dusk with a pelagic trawl.

The diurnal movement of DSL's has been observed in the Atlantic, the Indian Ocean and the Okhotsk Sea in the Pacific. The thickness and movement of the DSL's is considered by the Japanese to provide an important indication of the productivity of fishing grounds.

6.7 Fishing gear behaviour

It is possible to observe the performance of fishing gear by using either an echo sounder or a sonar. This is best achieved from a vessel other than that doing the fishing. A bottom trawl observed by a high frequency sector scanning sonar is shown in *Fig 69*

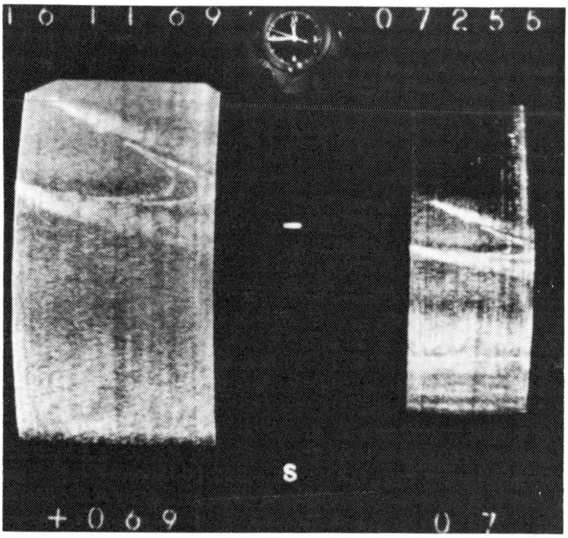

Fig 69 The front section of a bottom trawl as seen on the CRT display of a sector scanning sonar. The headline, groundrope and eddy trails of mud from the otter boards are clearly recognizable *(British Crown Copyright)*

Fig 70 shows the shape of a tuna longline in operation, obtained by a high frequency (200kHz) sounder. *Figs 71* and *72* show a purse seine net as recorded from a netsonde and *Fig 73* shows the use of a netsonde for 'aimed' midwater trawling.

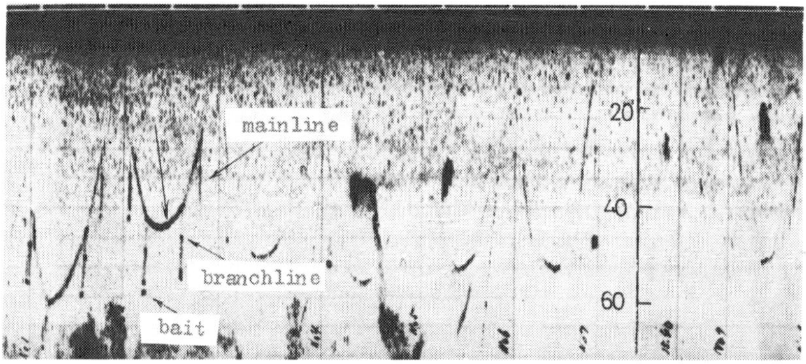

Fig 70 Echogram made by a 200kHz echo sounder, showing the shape of a tuna longline

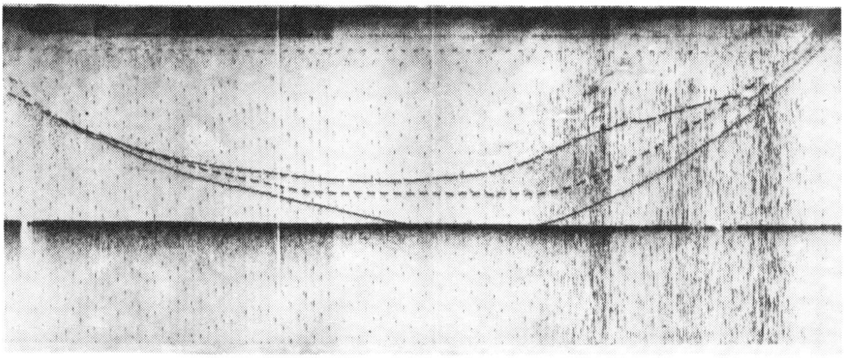

Fig 71 Sinking pattern of a purse seine, observed by a three-frequency netsonde device (40, 50 and 60kHz) and recorded on a 0-160m range echo sounder. Japan Sea

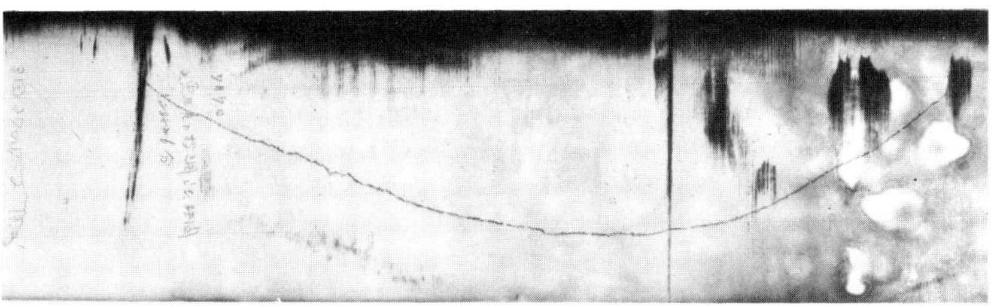

Fig 72 Netsonde echogram of a purse seine, made off Kushiro, Hokkaido. The fish are mackerel

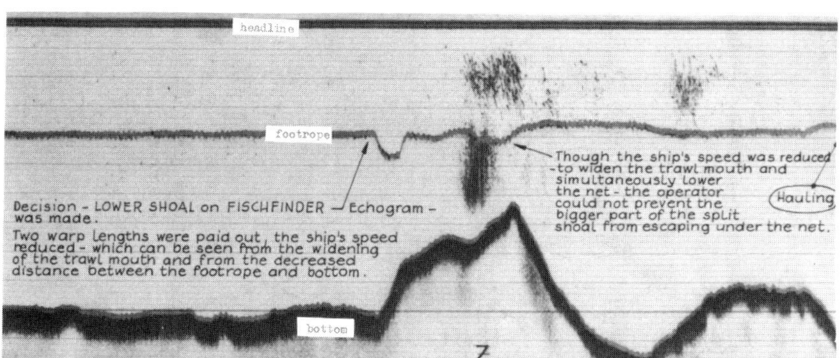

Fig 73 An example of 'aimed' midwater trawling combining ship's sounder (not shown) and netsonde observations for manoeuvring the net to hit a fish school. Note evasive reaction of fish

CHAPTER 7

DESCRIPTION AND USE OF SONAR

7.1 Main components

Sonar operates in much the same way as an echo sounder and has the same four main components (*Fig 74*), the transmitter, the transducer, the receiver and the recorder or display. The function and performance of each component is basically as described in Chapter 2, but as we shall see in the following chapter, with some modifications, due to the transmission being more or less horizontal and to the consequent longer ranges involved.

7.1.1 Transmitter

As in the case of the vertical echo sounder, the transmitter puts pulses of electrical energy to the transducer, but in order to obtain the required longer range, there are differences in frequency, power and pulse length.

Lower frequencies are used because they travel further in water (Chapter 1). Common sonar frequencies range from 20 to 40kHz. The power output is increased up to 10kW as compared with about 1kW for an average echo sounder. To get such high power into the water, the pulse lengths are increased up to 40ms. This greatly reduces fish discrimination, but this is not so important for long range searching for fish schools. The pulse length should be shortened as the range decreases and most sonars are equipped for this.

7.1.2 Transducer

The properties of transducers were discussed under *2.1.2*. Because the transmission is more or less horizontal, pitching and rolling of the vessel is important and a sufficiently wide beam or a stabilized narrow beam are required. The sonar transducer must be able to move in azimuth and should also be tiltable, usually from about + 5°, *ie* slightly towards the sea surface, to –90°, *ie* vertical down to the sea bed. In order to achieve this, the transducer is mounted in gimbals (*Fig 75*) and in order to avoid the turbulent layer and aerated water along the hull of the ship, it is lowered to about 1m below the ship's bottom when in operation

The transducer assembly must be able to withstand considerable mechanical stress when the vessel is under way and must be retractable when not in use, when docking, or in very shallow water; also to avoid fouling fishing gear, such as a purse seine. Some larger sets have a built-in safety feature which causes the transducer to be retracted automatically when the set is switched off.

The control for the transducer is usually mounted, with the recorder, in the wheelhouse console. Most sonars are able to train the transducer through 360° and to indicate its position by means of a pointer on the console (*Fig 76*). Similar controls and an indicator are provided for the angle of tilt.

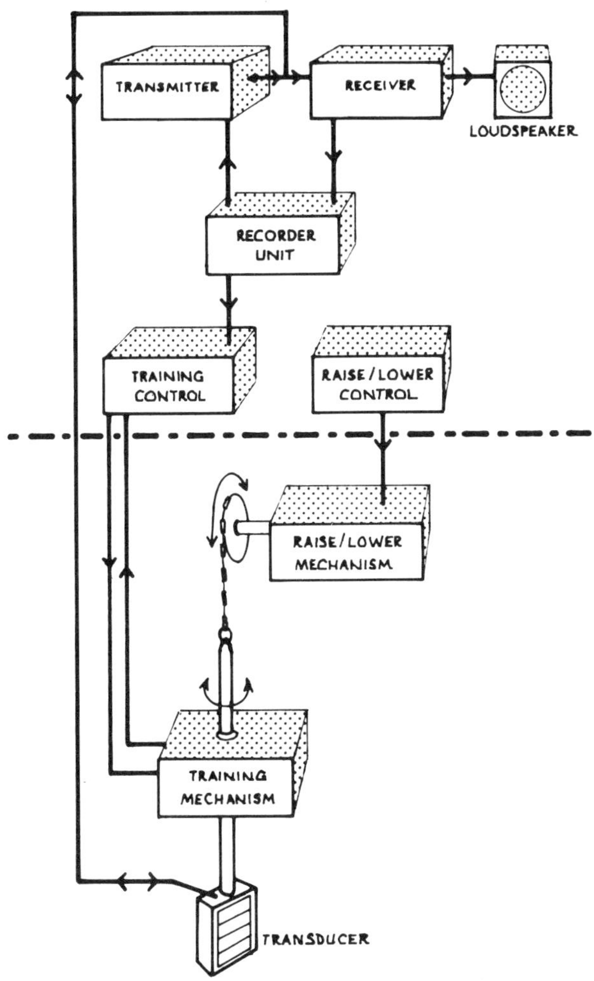

Fig 74 Block diagram of the main components of a sonar installation

The positioning of the transducer in the hull of the vessel is not a simple problem and for this, most manufacturers offer a specialized service. After installation, comprehensive checks should be made to ensure that the equipment is operating correctly. These checks can indicate the maximum

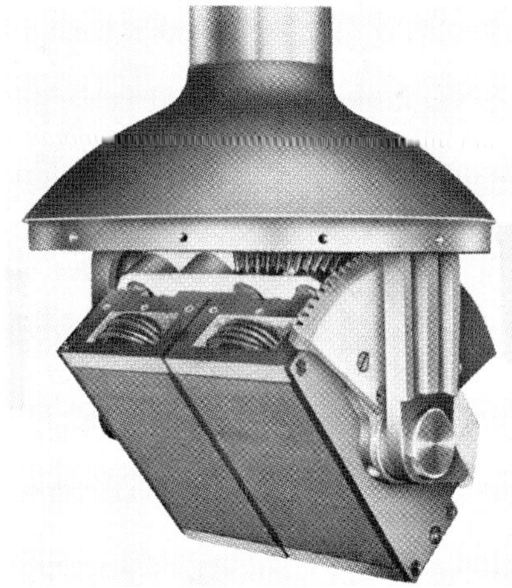

Fig 75 Tiltable sonar transducer arrangement. The spherical cover has been removed

vessel speeds for efficient searching and, if applicable, the optimum propeller pitch (*Fig 77*). For ships operating at high speeds (above 12 knots) it is necessary to have a streamlined cover for the transducer to reduce noise from turbulence around the transducer face (*Fig 78*).

7.1.3 Receiver

The sonar receiver is similar in operation to that described under *2.1.3*. The very weak signals that are returned from fish to the transducer are amplified to a sufficient magnitude to mark a paper chart, operate a CRT and produce an audible indication from a loudspeaker.

7.1.4 Recorder

The paper recorders are in principle the same as for echo sounders (*2.1.4*). Since it is not the depth but the more or less horizontal distance of an object from the vessel which is shown, the display of an approaching fish school appears as a slanting line on the sonargram as the distance changes. A display which is better able to indicate distance and direction is the plan position indication (PPI, see *9.4*) on a CRT, which is similar to a very slow radar. The ship is the centre of the scope and the sonar beam can be seen travelling away from the ship on the bearing selected. If a target is detected the echo is shown on the screen, giving an indication of its range and bearing. The approximate depth of the target can be calculated from the distance and the angle of tilt (but remember Chapter 1, *Fig 3* and *8.1*). A paper recorder may be combined with a CRT for closer study.

The audible indication of the echoes by loudspeaker is quite essential for the identification of various targets. An experienced operator can distinguish between different types of echoes better by the sound than from a paper record or CRT display (*8.3*).

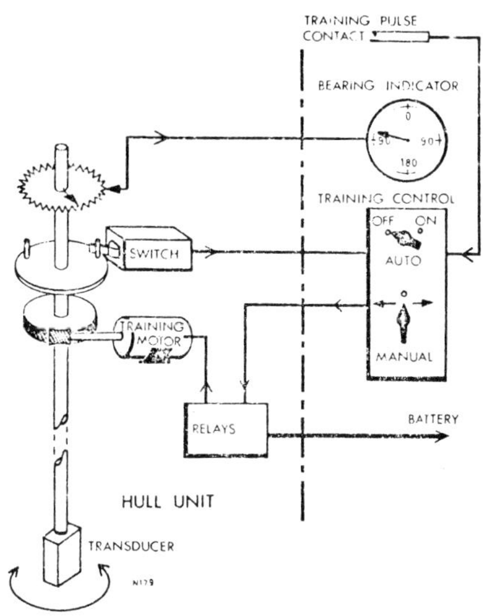

Fig 76 Block diagram of the training system

7.2 Search Methods

Sonar is important for fish searching, particularly for purse seining and midwater trawling. It is also quite useful for assessing bottom conditions and configuration, which are of interest for bottom trawling and other bottom fishing techniques.

All sonars have manual tilt and training control. More elaborate ones have in addition automatic preset training programmes for the transducer, which may be by steps or continuous. In a stepped training system, the transducer is held in the transmitting position to allow reception of all the possible echoes from the range scale selected. The bearing is then altered by a step equal to about half the beam angle. The total time between two transmissions, or the pulse repetition frequency (PRF), depends on the selected range and should be adjusted to suit the actual obtainable sonar range. (Chapter 1, *Fig 3* and *8.1*). If this adjustment is not carried out properly, time will be wasted waiting for returns from distances from which no echoes can be expected and, as during this time the vessel will be moving, the area within the actual sonar range is covered less thoroughly than

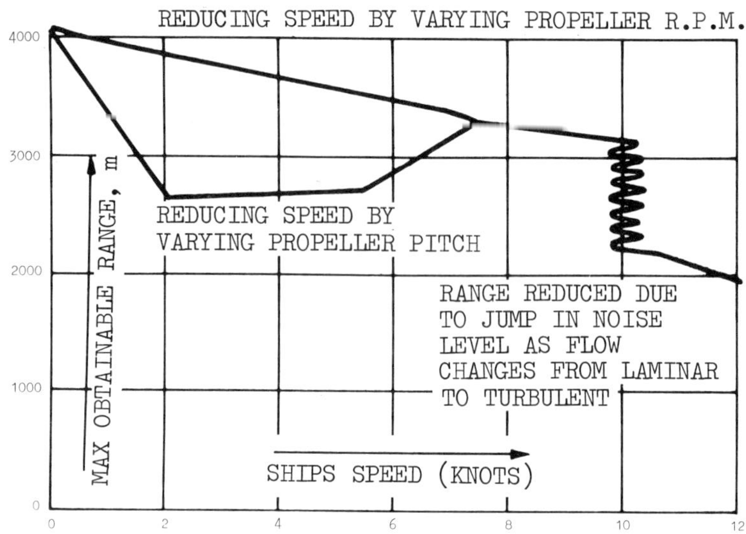

Fig 77 Example of the effect of adjusting ship's speed by changing propeller pitch or by varying engine revolutions (rpm). By varying the pitch, more noise is created and a loss of range results. Note that a significant range reduction occurs at 10 knots. For a long range search, 9 knots gives much better results

possible. In other words, by using a longer range scale than the actual detection range set by the sonar conditions, there will be unnecessarily large areas around the ship uncovered or missed, where fish schools or other targets might escape detection.

There are three common automatic search patterns (*Fig 79*):

Side to bow. The search is always made towards the bow and, on reaching the ahead bearing or a few degrees beyond it, the transducer is trained rapidly to the start bearing on the other side.

Side to side. The transducer is trained from one side through the ahead bearing to the other side and back again.

Side to side and back. This programme differs from 'side to side' in that the transducer is returned quickly to the start position as soon as both sides have been searched.

In open water the side to bow search is the most efficient as indicated in *Fig 79*. When searching on shelving sea beds, it is better to steam close along the shore and search mainly out into the deeper water.

When searching, the tilt control should be set to the angle which allows the main part of the sound beam to cover the volume of water where fish

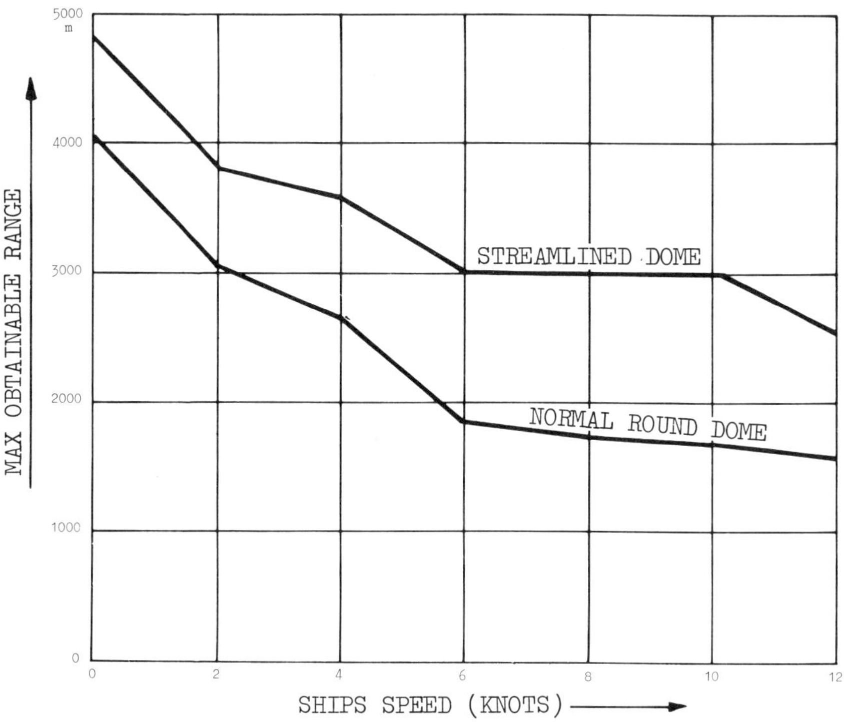

Fig 78 Example of the advantages of a streamlined dome over a normal one, illustrating the large decrease of range as the ship's speed is increased. Note that this diagram refers to a less powerful sonar than that illustrated in *Fig 77*. Vertical axis shows maximum obtainable range in metres

are likely to be found. If a promising echo is received from a long range, the ship's course is altered to the bearing of the target and the transducer is trained continuously over a narrow sector keeping the target straight ahead. When the range is reduced to about 400m to 500m, the ship's speed may be reduced. Automatic searching control should then be changed to manual

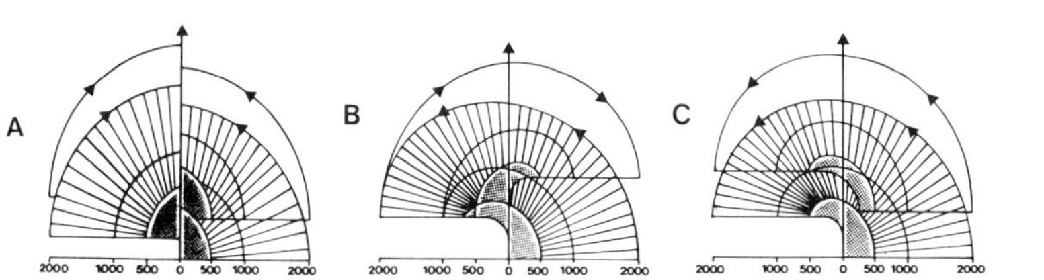

Fig 79 Coverage of three common search patterns at 10 knots ship's speed. A, side to bow; B, side to side; C, side to side and back. Horizontal scale in metres

control for the evaluation phase. Some indication of the density of a fish school can be obtained by approaching to within about 200m and reducing first the pulse length, then the receiver gain and lastly the transmitted power. If the echo fades rapidly for small reductions, then the school is not dense. With dense schools, the echo display is maintained even with the shortest pulse length, low receiver gain and reduced transmitted power.

Target movement may be detected by Doppler shift (*8.3*), although a fish school is generally a slow-moving target and therefore the small frequency changes are difficult to observe. The changes in range and bearing from a stopped ship is another way of observing school movements and a third method is to circle the school at a distance of approximately 200m and measure the change in distance between the echoes from the propeller wake and the school.

Having decided to enter the catching phase the fishing gear is shot away and the sonar is used for a close range observation and navigation. The relative movement between fish schools and the ship at close range can be rapid and this means that the transducer must be trained quickly back and forth across the target to maintain contact. The highest pulse repetition frequency, *ie* the shortest possible range, should be used.

When midwater trawling, the trawl can be positioned approximately for the school on the basis of a depth calculated from the indicated distance and the transducer tilt angle. Contact is then maintained and the trawler will usually pass over the school so that a more accurate depth measurement can be obtained from the ship's echo sounder. Knowing this depth, the vessel's speed can rapidly be altered, or the warp length varied, in order to position the trawl in the right depth to catch the maximum quantity of fish. Depending on species or seasons of the year, fish may be scared by the noise of the ship and allowance for this must be made when setting the trawl depth. If the school is close to the surface, it is difficult to retain sonar contact astern through the aerated propeller wake. To facilitate correct depth adjustments of the midwater trawl, multi-netsonde equipment with up to six transducers on the headline has been developed (*5.4*).

CHAPTER 8

MORE ABOUT THE FUNCTION OF SONAR

The fact that the transmitted beam operates more or less horizontally rather than vertically introduces a number of differences as compared with normal echo sounding. The three main factors to be considered are beam bending, reverberation and the Doppler effect.

8.1 Beam bending

This effect was introduced in Chapter 1 and results principally from the change of temperature (or salinity) with depth. Sound travels faster in warm water than in cold, hence the beam bends towards the cold water. If we consider a horizontal transmission from a transducer, the wave front will have a height dimension which will depend upon the beam width. Where there is a temperature gradient in the sea, the top of this wave front will travel at a different speed to the bottom. In *Fig 80*, the water at the top of the front is warmer than that at the bottom and the beam will be bent downwards. The converse will happen if the water at the top is colder.

The bending of beams results in a false indication of range and in particular, depth. It also creates shadow zones, where fish may escape detection. These facts should be remembered when operating sonar in different seas and at different times of the year. In winter the sea is generally cold at the surface and warmer with depth. This bends the beam upwards, giving rise to the type of pattern shown in *Fig 81a*. In summer, the upper layers are warm, getting colder with depth. This gives rise to the condition shown in *Fig 81b*. Sharp downward bending of the beam limits the sonar range and brings in interference from bottom echoes. Hence sonar conditions are particularly poor during hot, calm seasons.

8.2 Reverberation

Reverberation, discussed under *3.4*, is the back scattering of the transmitted pulse by particles and irregularities both in the water and on the surface or the bottom. When the sonar beam hits the sea bed, bottom reverberation is produced. If the bottom is rough, a large amount of energy can be returned and can completely mask any echo from targets within the beam after the first return from the bottom. As can be seen from *Fig 82* this restricts the use of the sonar for the detection of demersal or bottom fish to be caught with the normal bottom trawls (headline height about 2 to 3m) which will be within the shaded area. The reverberation intensity depends very much on the angle of the beam with the bottom (the 'grazing' angle), getting stronger with increasing inclination. It is weakest when parallel to the bottom. This feature helps when using the sonar in shallow water and also explains why it is mainly used for seeking midwater or pelagic fish.

There are two ways to reduce reverberation, by use of a narrow beam, or a short transmission pulse.

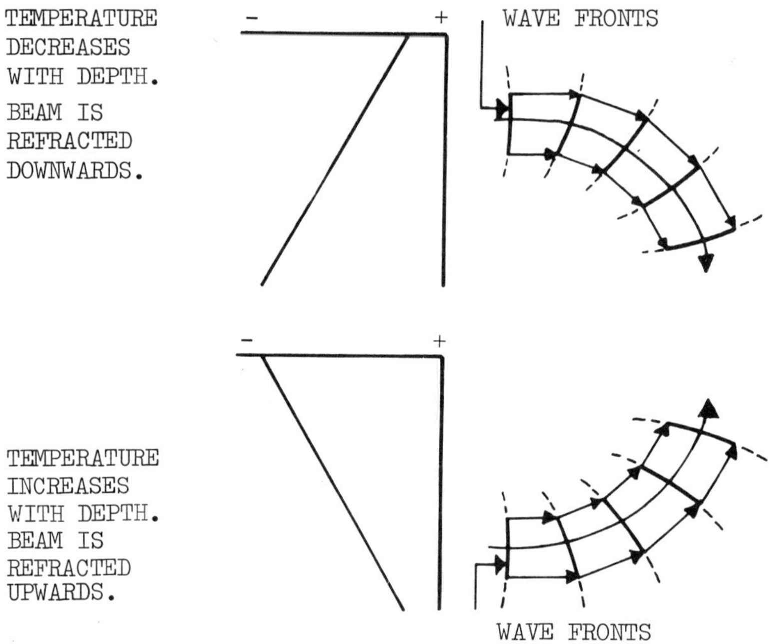

Fig 80 Refraction of a sonar beam's sound wave front with temperature

By reducing the beam width, the area of bottom in the beam is smaller, thus reducing bottom reverberation. The number of midwater scatterers (volume reverberation) is also reduced. A limit to this reduction of the beam width is imposed by the pitch and roll of the vessel, due to which small targets will be within the beam only for part of the time thus reducing the detection capability of the sonar. This can be overcome by stabilizing the beam but this is expensive.

The interval of range from which reverberation is obtained is related to approximately one half of the pulse length transmitted. Thus, shortening the pulse can be an effective method for improving the detection of an echo from a target masked by reverberation. The limit to this reduction is when the pulse length equals the range length of the target. Beyond this, the target echo is reduced as well as the reverberation and the signal/

reverberation ratio remains constant. However, this would occur only with very small schools. As in echo sounding, the short pulse is used to resolve targets which are close together.

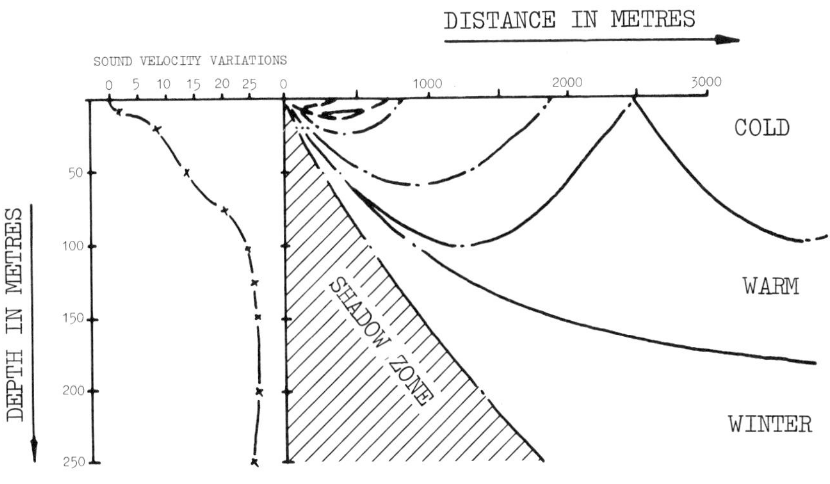

Fig 81a Sound propagation when sound velocity increases with depth

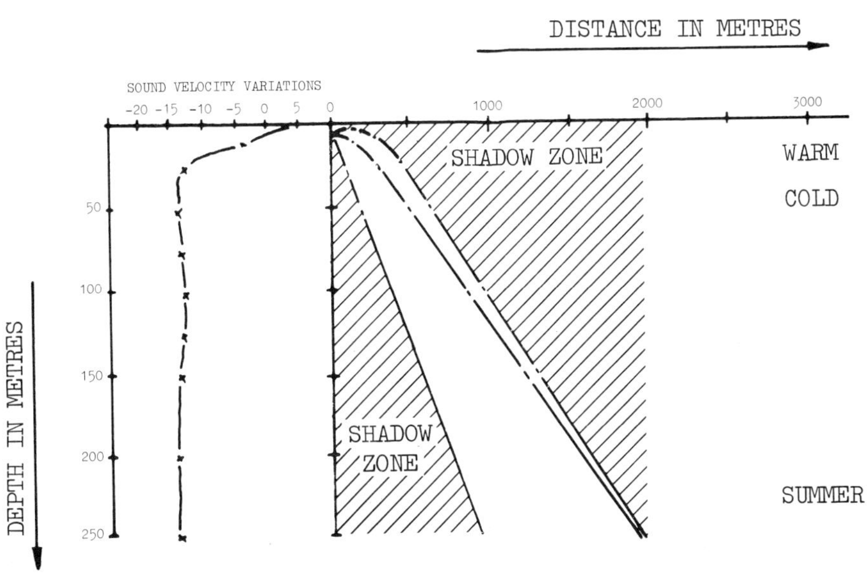

Fig 81b Sound propagation when sound velocity decreases with increasing depth

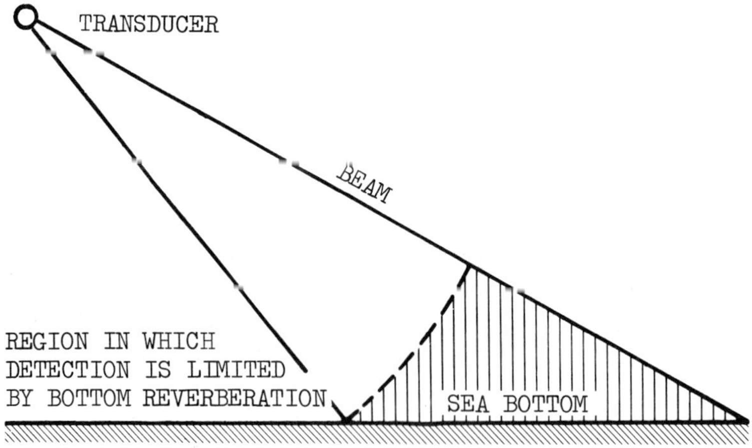

Fig 82 Bottom reverberation

8.3 Doppler effect

In addition to the wet or dry paper recorders and the CRT or fishlupe as used for echo sounder display, the sonar also has a loudspeaker relaying the sound quality of echoes. An experienced operator uses the sound to distinguish fish echoes from surface or bottom echoes and also from reverberation and background noise.

If the listener were to move towards the gong in *Fig 1*, then the rate at which the sound waves would be received would increase, *ie* the frequency received would be increased above that transmitted. In the same way the frequency is decreased when the listener moves away from the sound source. This frequency change is called the Doppler shift. It should be noted that the Doppler shift (f_d) from echoes occurs both over the transmission path and over the return path and is given by:

$$f_d = 2f_t\frac{V}{c}$$

Where f_t = frequency transmitted;
c = speed of sound in water;
V = relative velocities of vessel and target in the direction of the sound beam.

Targets can be detected by listening for Doppler shift only if they are moving with some speed, which tends to limit its usefulness to whale and possibly tuna detection. If the target is to be detected by Doppler, then the earlier recommendation of using a short pulse to limit reverberation cannot be adopted. The longer pulse produces a purer tone from the target and therefore makes the Doppler shift more discernible to the ear.

With suitable instruments the sound frequency can be measured with much higher accuracy than by the human ear and the Doppler shift is therefore utilized for accurate measurement of ship's speed over the bottom and/or through water (Doppler log).

CHAPTER 9

DIFFERENT TYPES OF SONARS

The main components of a sonar installation are shown in *Fig 74*. The units above the dotted line are usually found in the control cabinet and those below the line are installed in the bottom of the ship. The displays are normally in the form of a paper recorder and a loudspeaker, but CRT displays are also used in some sonars, either instead of recording (in smaller units) or in addition to recording to provide a more instructive 'panoramic' presentation with regard to distance and bearing of targets.

On simple sonar units for smaller vessels, the transducer may actually be hand operated for lowering and lifting and for training and tilting. Larger vessels and more elaborate sonars use a remote control of the transducer. Training is either manual or automatic through predetermined programmes (*7.2* and *Fig 79*). Remote control is from a bridge unit where instruments indicate the transducer bearing and angle of tilt.

9.1 Short range, hand operated training

Such sonars are suitable for small boats down to approximately 5m in length. The working range is normally about 500m and the equipment is constructed for low speeds of less than 8 knots. As an example, this type may be installed in the small fishing-master boat used in certain purse seine fisheries. Having first detected the fish school from the main boat, this small boat is launched to maintain contact, report on the speed and direction of movement of the school and direct the setting of the net.

9.2 Medium range, automatic training

The second group of sonars operates to a maximum range of approximately 1 500m. The transducer is remotely controlled. These sets can be used up to speeds of 12 knots but, as the transducer has no streamlined dome protection, noise can reduce the working range at speeds above 9 knots. These sets are used in medium size vessels of about 50 to 200 tons.

9.3 Long range, automatic

This group has remote control of transducer tilt from horizontal to vertical, automatic search programmes and manual training through 360°. The

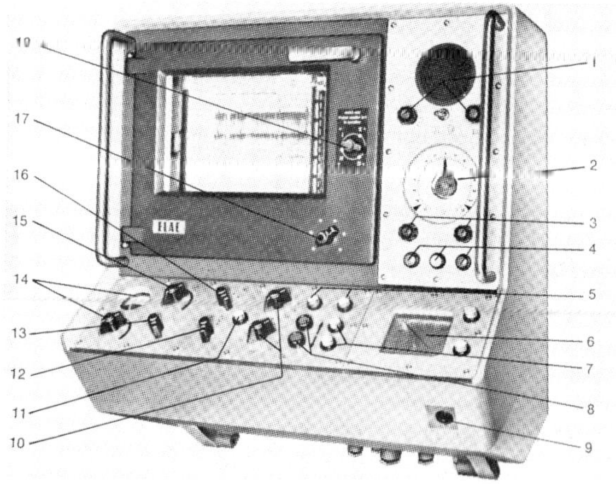

1 Loudspeaker with adjusters
2 Transducer bearing and search sector indicator
3 Search sector adjusters
4 Hand-training controls
5 On and Off push buttons
6 Tilt indicator and controls
7 Transducer Up and Down controls
8 Up and Down pilot lamps
9 Plug-in for headset
10 Gain controls (H + V)
11 Push-button marker
12 Operation selector switch
13 Pulse Lengths switch
14 Voltmeter and regulator
15 Dimmer
16 Transducer selector switch
17 Paper speed adjuster
18 Scale selector switch

Fig 83 Operating panel and recorder unit of a fishing sonar with automatic transducer control

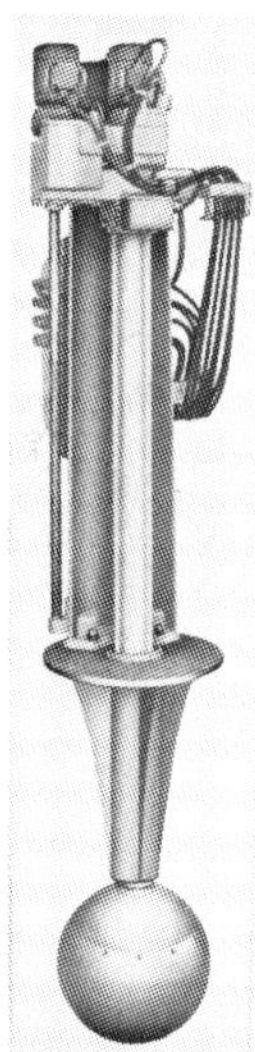

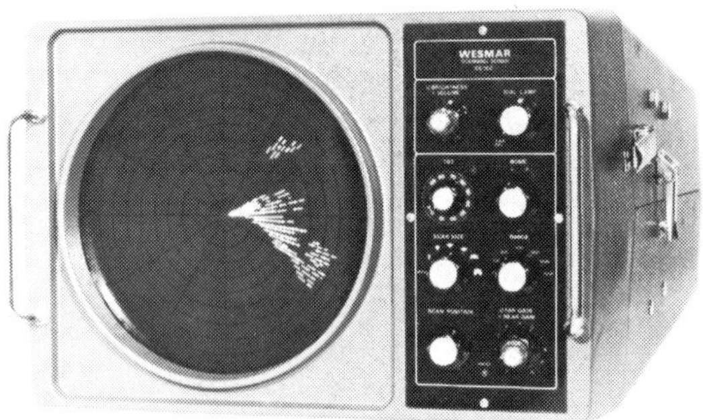

Fig 84 *(Left)* Transducer arrangement with hoist/sweep gear
Fig 85 *(Above)* Example of a short range, high frequency sonar with PPI/acoustic display and gravity-stabilized transducer

transducer may be protected by a streamlined dome. Maximum operational speeds of about 15 knots and ranges of 3 000m to 4 000m are typical of this group. *Fig 83* is an example of the recorder and control panel and *Fig 84* shows a transducer arrangement with hoist and sweep gear. The transducer is automatically retracted into a protecting sea chest when the set is switched off.

9.4 Plan Position Indicator (PPI) displays

The cathode ray tube (CRT) display on some sonars is similar to that of radar. However, the speed of sound in water is comparatively low and the scanning rate for full 360° coverage is consequently also low. For example on a range of 3 000m it takes 4 seconds for all the echoes to be received before the transducer can be stepped on, say 5°. To cover 360° in this way

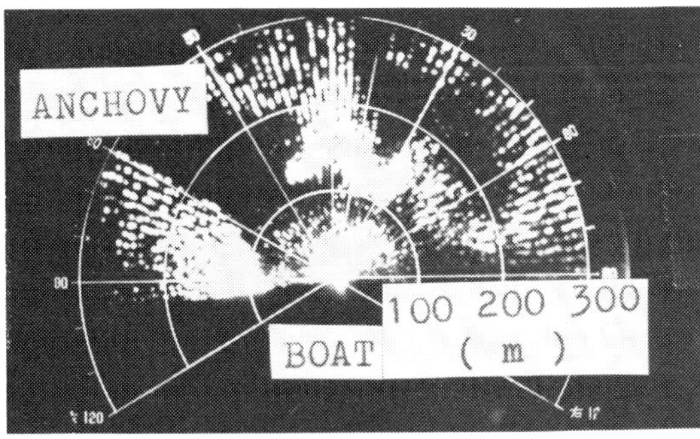

Fig 86 A characteristic short range sonar PPI display

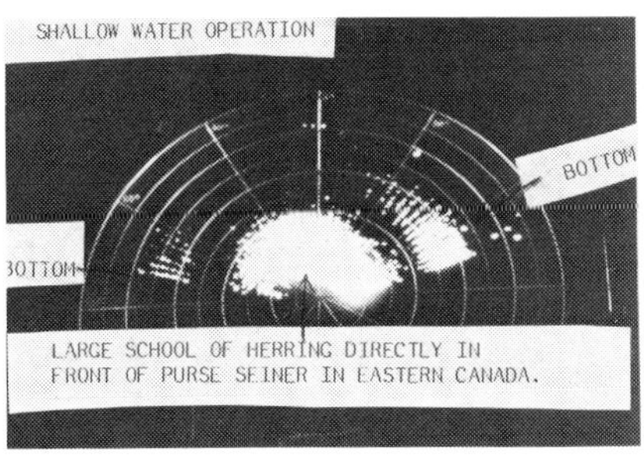

Fig 87 Another typical short range sonar PPI display

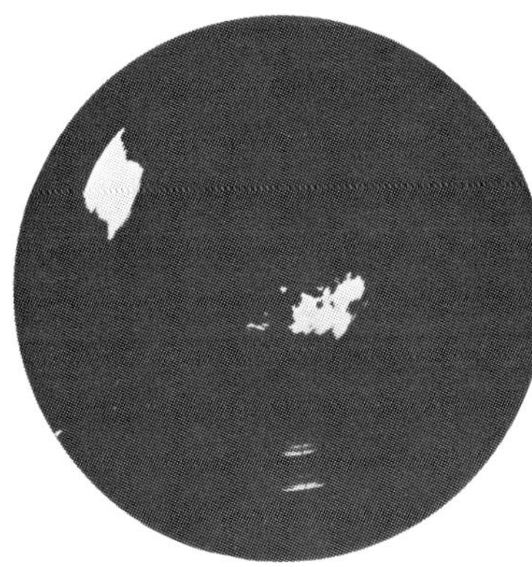

Fig 88 This photograph of fish schools as seen on the CRT/PPI display was taken with the range scale set at 1,000m. The school on the port bow is mackerel and is at a range between 650m and 850m. This resulted in a catch of 200 tons of fish, while the school nearer to the vessel gave 60 tons. The effect of wake turbulence can be seen astern, but it has no adverse effect on the echoes from the two schools

to obtain a full panorama would take about 5 minutes. Consequently, this type of panoramic, or rather sector display is more suitable for short ranges of up to 500m. The transducer is usually trained continuously within the smallest acceptable sector, and a CRT with a long persistence is selected to retain the image and reduce the flashing effect as further information is received. There are units with gravity-stabilized transducer, using medium to high frequencies with limited range (*Fig 85*).

To reduce the scanning time and thereby obtain longer ranges of up to 4 000m for sector display, higher powered sonars are produced with special multi-element transducers. These elements can be switched electronically to allow either omni-directional transmission, *ie* covering 360° simultaneously, or transmission over selected areas, *eg* 210°, 180° or 90°. Following transmission, the transducer elements are again switched to obtain pre-formed narrow beams, *eg* 36, each of 10°, and these are electronically scanned for echoes. In the more sophisticated and expensive of these 'Panoramic' sonar sets, the information received is stored in an electronic memory from which it is recovered and displayed to obtain a steady, flicker free CRT picture. By adopting this method of using a memory storage, special signal processing techniques can be incorporated in order to reject interference from other sonars, remove false echoes and improve the signal-to-noise ratio.

Figs 86 and *87* are examples of CRT displays from small, high frequency sonars of 150kHz to 200kHz. *Fig 88* is taken from a powerful 30kHz set.

CHAPTER 10

SONARGRAMS AND THEIR INTERPRETATION

Fig 89 illustrates how the sonargram from fish and sea bed is produced as the vessel moves forward, so reducing the distances to targets. This is an ideal situation, as is the sonargram from a herring school shown in *Fig 90*. The traces are seen as sharp, distinct lines and the loudspeaker would give strong sound signals, easily distinguished by an experienced operator. The first mark to the left on the chart is the transmission and the second mark parallel to it is the start of reverberation from the bottom, generated by the edge of the beam (*8.2*). *Fig 91* shows the traces from three schools of mackerel both on sonar and on echo sounder.

When fishing in conjunction with sonar, a rational search pattern should be employed (*7.2*) and the loudspeaker and other displays observed for the first sign of a contact with a fish school. When this is detected, the transducer is held on to the contact, either manually or by limiting the automatic scanning sector, and the vessel is turned on to the target bearing. As the range closes, the scale of the sonar should be reduced. The effect of range changing after making first contact is illustrated in *Fig 92*. In *Fig 93* the first contacts are made on the 2 500m range and the strong fish school traces show clearly within the random marks generated by bottom reverberation. On the 500m scale the trace is shown rising through the first return from bottom reverberation, thus showing it to be from a midwater fish school rather than from a pinnacle on the sea bed. *Fig 94* is an example of approaching a tuna school from a distance of 2 500m, switching to the 1 250m range when the school is at 1 000m and finally closing to 300m on the 500m scale before moving away from the school and switching again to the 1 250m range.

Figs 95 and *96* are further examples of characteristic sonargrams. *Fig 96*, taken on board the Norwegian research vessel *GO Sars*, shows sonar contact with a school of herring. This again is a good illustration of a trace of a fish school in midwater passing through the front edge of the sea-bed trace and coming clear of it when observed on the short range setting (1). The school is then observed on the echo sounder (2) and, finally, the trawl is adjusted to the correct depth to catch the fish as shown on the netsonde echogram (3).

Reverberation is always present on the sonargram and the extent to which it covers its width can indicate the maximum working range of the equipment when operating in the open sea and with shallow tilt angles on

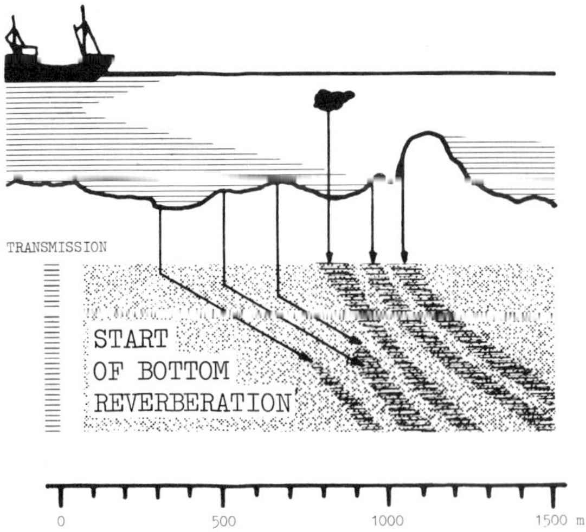

Fig 89 *(Left)* Scheme of a sonargram showing bottom irregularities and a fish school. Additional acoustic display is essential in order to recognize the fish school echo

Fig 90 *(Right)* Sonar trace of a herring school being approached

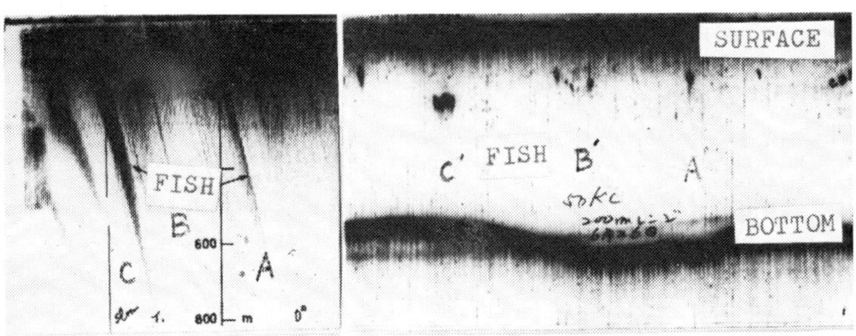

Fig 91 Echo traces as shown by sonar (left) and by an echo sounder (right). Corresponding traces are marked A, A'; B, B'; and C, C'

the transducer. Further information on the actual working range can be obtained from echoes returning from targets below or between different temperature layers or internal waves.

With small angles of tilt, reverberation from the surface appears at distances from zero to a few hundred metres all round the horizon. Its strength may vary seasonally. Surface waves may give the appearance of a distinct target but can be recognized because they soon break up.

Echoes from propeller wake may be a considerable nuisance when operating in a fleet of vessels and can mask targets. This can be overcome by manoeuvring appropriately around other vessels.

Interference from other sonars or echo sounders is also a nuisance, but can easily be recognized, as this appears as a characteristic and rather regular pattern of traces on the chart.

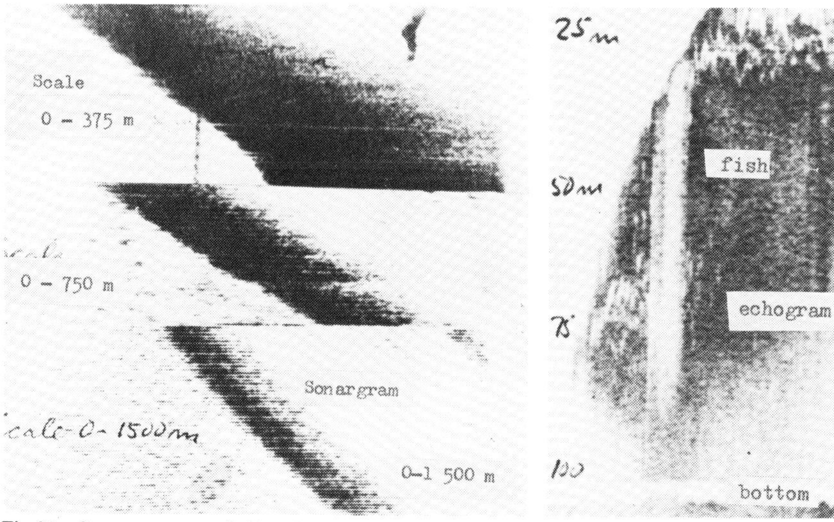

Fig 92 Sonar contact (left) with a huge school of anchoveta, in daylight, during the September inshore migration off Argentina. Note reduction of sonar range while approaching the school, from 1,500m to 750m and then to 375m. Corresponding trace from ship's sounder is shown right

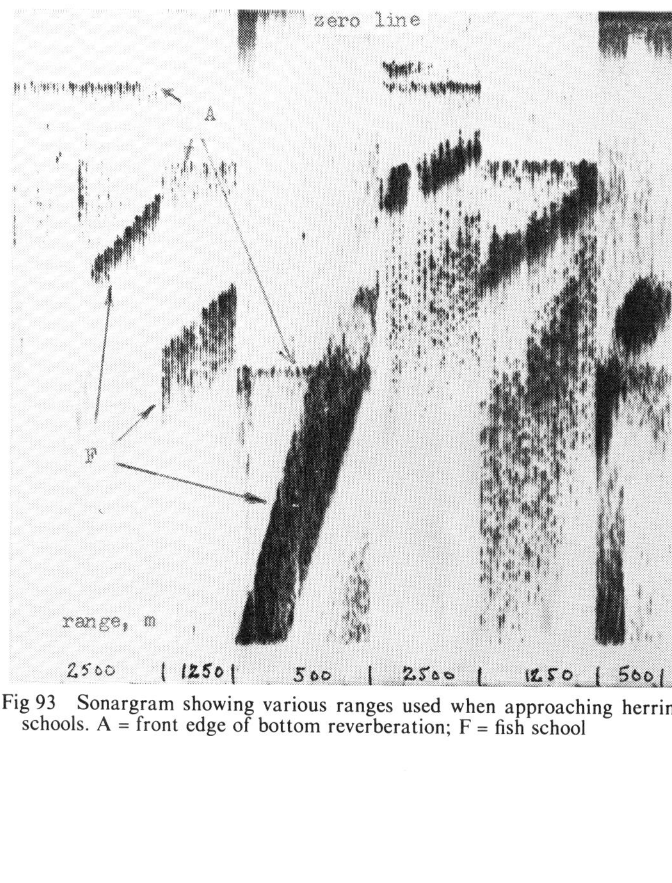

Fig 93 Sonargram showing various ranges used when approaching herring schools. A = front edge of bottom reverberation; F = fish school

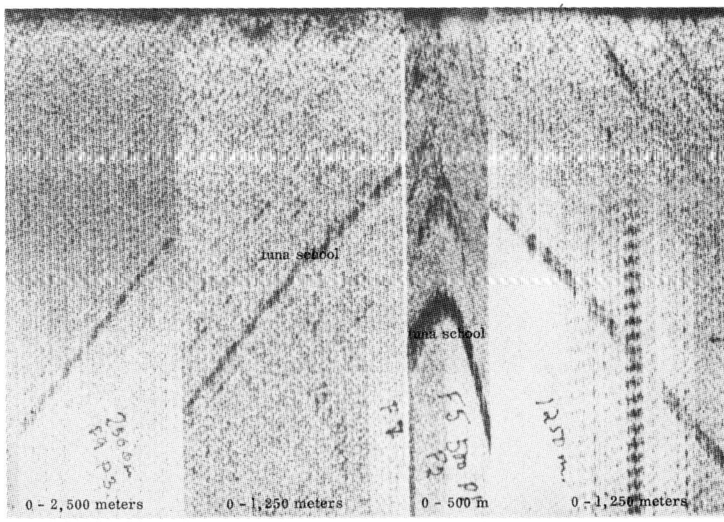

Fig 94 Sonargram showing approach towards a school of tuna, passing it at a distance of 300m and then leaving it behind. By training the transducer from forward to sideways and then backwards, the school is kept continuously in the sound beam

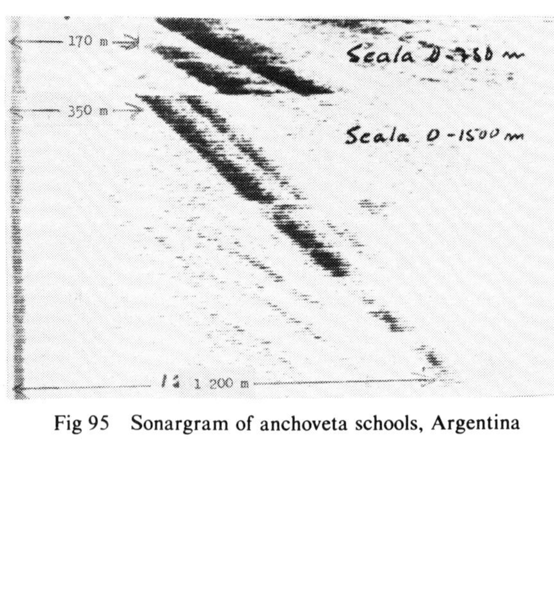

Fig 95 Sonargram of anchoveta schools, Argentina

Echoes from sizeable objects on the sea bed such as wrecks, rocks *etc,* can look similar to a fish school. The best method to establish such a target's identity is to approach it in order to see whether it merges with the bottom reverberation or crosses through the front edge to be shown clearly in

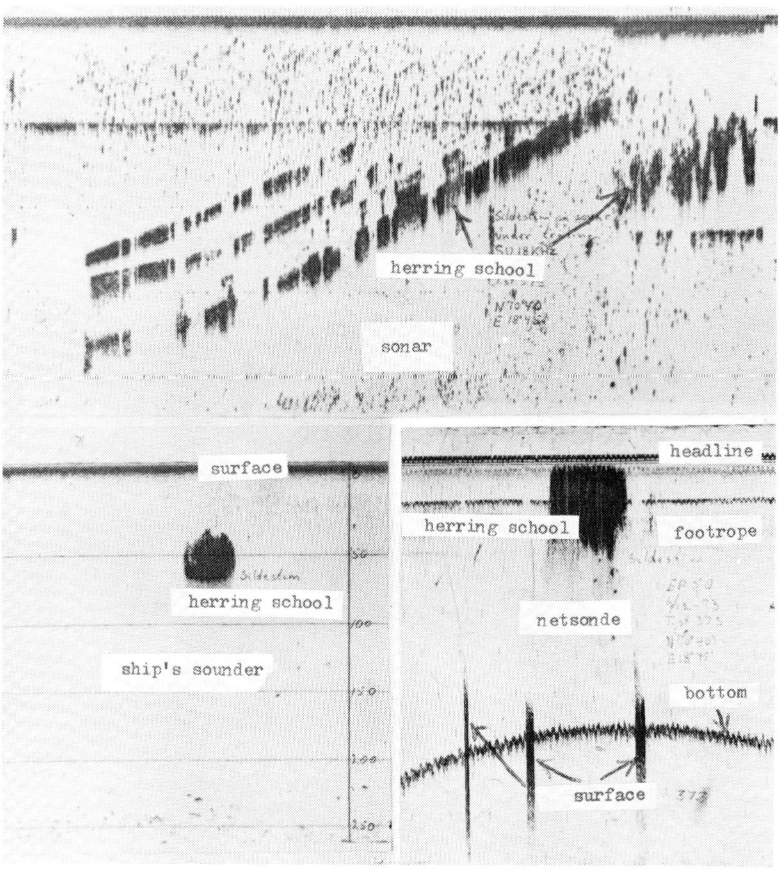

Fig 96 Records of sonar, echo sounder and netsonde illustrating how to catch a school of herring by 'aimed' midwater trawling. After making sonar contact with a school of herring (above) the vessel is manoeuvred to pass directly over the school to assess exact depth on ship's echogram (bottom left). The trawl is adjusted to the correct depth, and the herring school enters the trawl (see netsonde echogram, bottom right). By switching from the downward netsonde transducer to upward sounding, the distance of the headline to the surface is obtained directly

midwater. Echoes from rock on the sea bed are shown in *Fig 97*. Note that there is little return of signal from flat sandy bottom, particularly at low angles of incidence, (*11.6*).

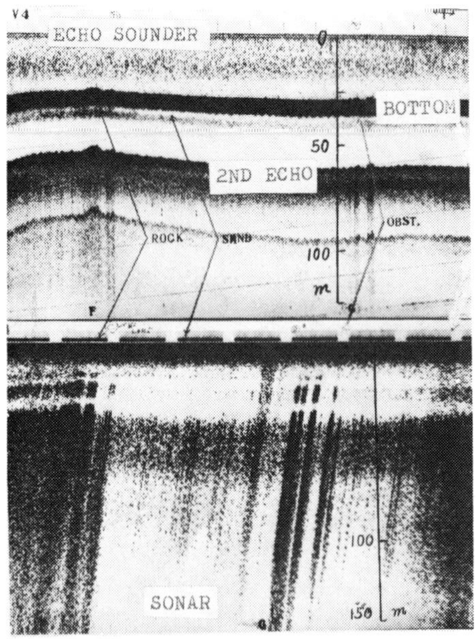

Fig 97 Echoes from rock and sandy bottom, as recorded by echo sounder (top) and sonar (below)

CHAPTER 11

FISHING WITH ECHO SOUNDER AND SONAR

11.1 Bottom (demersal) fishing

The bottom trawl is designed to be towed over the sea bed and may have rollers (bobbins) on the groundrope to improve passage over irregularities and obstacles on the bottom. The size and weight of these rollers depends on the type of bottom expected. The main and most common fishing aid for this catching method is the echo sounder and we have seen that techniques such as whiteline or greyline and scale expansion assist in the detection and identification of the fish being sought. The other obvious task for the echo sounder is to detect any dangerous features of the sea bed which may cause hook-up of, or damage to, the trawl leading to possible loss of money, fishing time and catch. Early warning enables alterations of course and other suitable measures to be taken in order to avoid such incidents.

To avoid damage the user must, for instance, be alert to the onset of hard ground (*6.1*). *Fig 98* illustrates rocky patches on the sea bed as displayed on an echo sounder using a wide beam transducer. Strong echoes from the power transmitted in the comparatively weak transducer side lobes are obtained from rocks. They produce the echo length and the oblique side lobe traces below the bottom trace, thus giving a clear warning of possible danger. With the rather low towing speeds and the length of warps the warning period provided by the ship's echo sounder is one to several minutes before the trawl reaches the rough patch, and this is sufficient time for hauling the gear or changing course, if it is thought that damage might result from continuing the tow. If the warning is short, then one of the hauling methods used when coming fast can be adopted, namely to stop the vessel while hauling so that, with the gear acting somewhat as a sea anchor, the vessel comes back over the gear. When the gear leaves the sea bed the vessel can be put ahead again (*11.6*).

Not all echo sounders are able to give side lobe echoes, which are also limited to small to moderate depths. The same sea bed seen in *Fig 98* is shown in *Fig 99*, this time using a narrow beam transducer with much reduced side lobes. Here the presence of the rocks is shown by the irregular profile, increased echo lengths and in particular the second bottom trace.

A similar risk of damage to bottom fishing gear is posed by isolated small wrecks or rock pinnacles on the sea bed. They are often less obvious, as is illustrated in the echogram in *Fig 100*. Although representing only a small

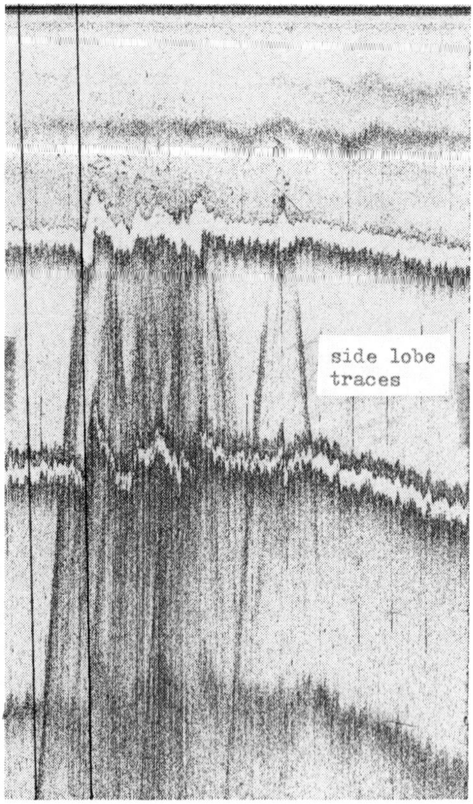

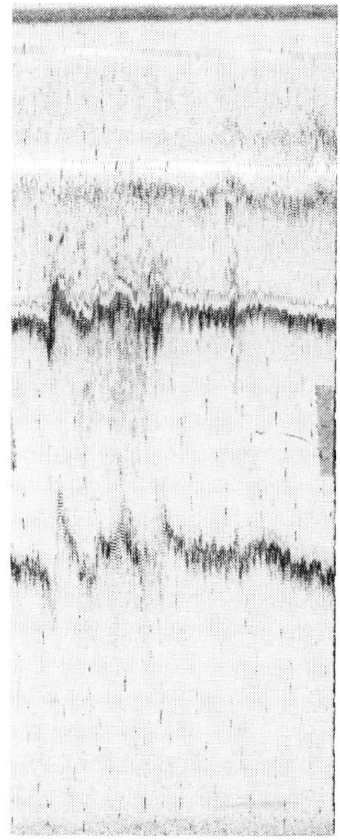

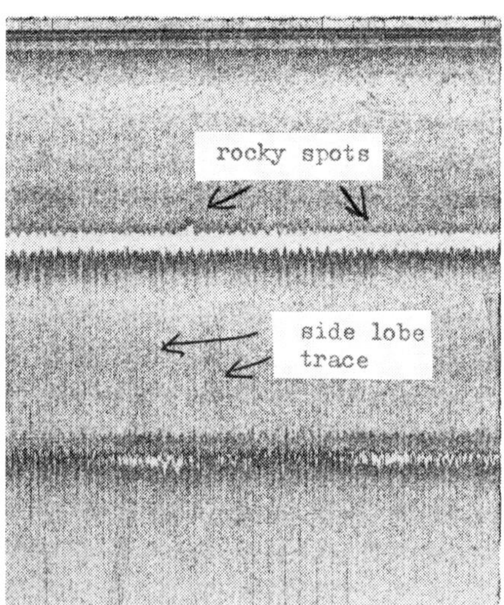

Fig 98 *(Above left)* Rocky patches on the sea bed clearly recorded with multiple echoes and 'echo length' by an echo sounder with wide beam transducer providing, in addition, oblique traces from side lobe echoes

Fig 99 *(Above)* The same rock patches as in *Fig 98*, but recorded with a narrow beam transducer with much reduced side lobes

Fig 100 *(Left)* Rock pinnacles recorded by a wide beam echo sounder. Note echo length due to side lobe echoes

variation in the bottom trace, early warning of the possible danger is again given to the attentive observer by the oblique trace from the side lobes; also by the stronger echo, giving whitelining in the second bottom trace and blacker marking of the third bottom trace.

The significance of assessing such bottom features by echo sounding, naturally applies to both Danish (anchor) or Scottish (fly dragging) bottom seining, or seine netting.

In many instances it has been found that bottom fish concentrate along a certain depth contour. To catch this fish it is essential to determine the extent and contour of slopes in order to maintain correct depth of a bottom trawl or to place other fishing gear in the right depth. It has to be appreciated that the steeper the slope the more a bottom trawl will tend to slide down it, and therefore the shallower the depth under the vessel has to be in order to hold the gear on the required contour.

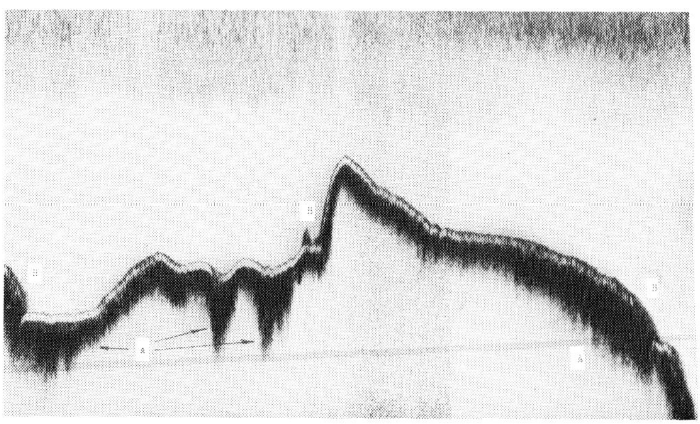

Fig 101 Echogram of trawling along a slope. Increased echo length (A) and fuzzy side lobe echoes with reduced whiteline (B) indicate particularly steep sections

Fig 101 is an echogram from a trawler towing along a steep slope, trying to keep the same depth. The irregularities of the actual depth under the trawler indicate that this was somewhat of a problem, probably due to a rather irregular contour of the slope requiring changes of course which were sometimes not enough or came too late. Characteristic steep slope effects are the long tails (echo length) of the bottom echo, the fuzziness of the whiteline and the traces above the whiteline, due to side lobe echoes from the slope arriving before the main beam echoes. These echo sounder indications help the vessel to maintain the desired depth. When trawling along very irregular slopes sonar can be used to assess the slope contour in advance, where the technique is essentially the same as described under *11.6*.

Fig 102 is an echogram showing an approach to a slope with increased echo length in the second half. Had this effect been due to hard ground then the second bottom echo trace would have been darker.

There are bottom fishing techniques such as longlining, handlining, gillnetting and pot fishing for which rough, rocky or sloping grounds, as well as rock pinnacles and wrecks, are advantageous rather than detrimental. Obviously echo sounding can assist very significantly in finding and evaluating such conditions for putting these gears in the best position (*11.4*).

Fig 102 Approach to a slope shown by increased echo length for the last part of the echogram. Care must be taken to distinguish this effect from hard bottom, which would be indicated by a stronger double echo

11.2 Midwater (pelagic) trawling

Midwater or pelagic trawling entails trawling in the pelagic or whole water volume, from just off the bottom to just below the surface. While bottom trawling may well yield fish from very close to the ground, where they are not recognizable by echo sounding, a midwater trawl has always to be 'aimed' at fish in reasonably dense concentrations (schools, layers), and echo sounding, sonar and netsonde are virtually indispensable for successful operation. Echo sounding alone is sufficient only under very favourable conditions, such as large schools, a large number of schools, compact layers in more or less uniform depth, shallow water, fish near the surface or the bottom and so on.

The 'aimed' trawling technique involves searching for, finding and evaluating schools with the sonar; correctly navigating the vessel towards

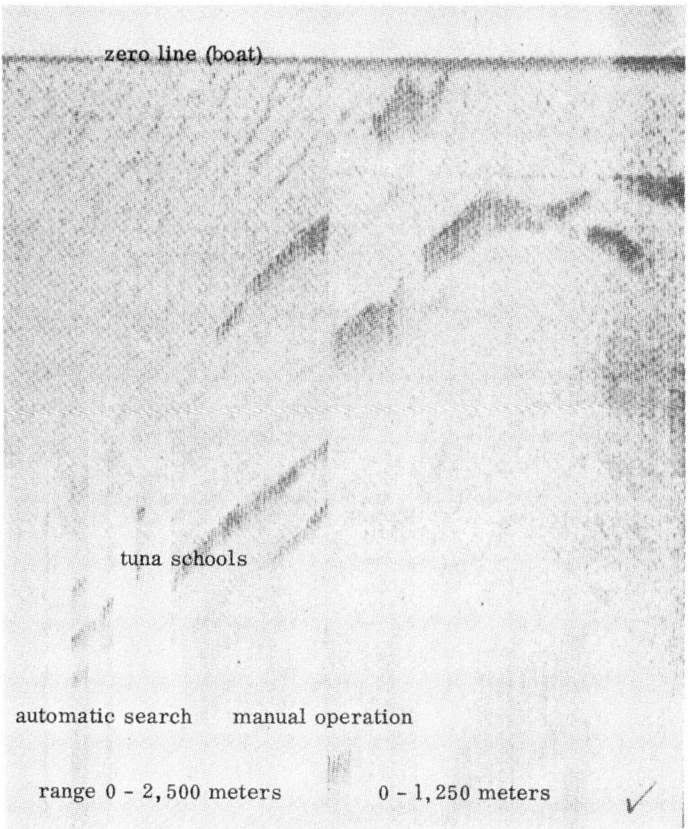

Fig 103 Sonargram showing tuna traces during searching

the selected school; establishing the accurate depth and configuration of the school by echo sounding and steering the trawl to the required depth, as indicated by the netsonde. The search patterns were discussed under *7.2* and typical search sonargrams are shown in *Figs 92–95* and in *Fig 103*.

Having made the contact, the school must be held by the sonar while the vessel approaches and eventually passes over it to assess depth, size and configuration. If found satisfactory the vessel turns round and shoots the trawl at a distance which will allow it to reach the required fishing depth.

The second approach is again sonar guided, if possible even after the vessel has passed the school, the better to ensure that the trawl will not miss the school sideways. If the trawl is already fishing, it must be noted that, subject to turning of the vessel, the trawl is mostly not astern, but follows in a

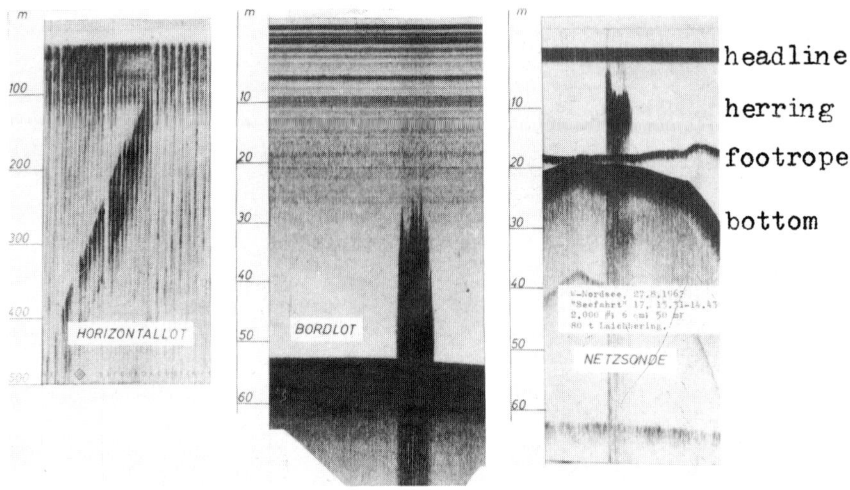

Fig 104 Sonar, ship's sounder and netsonde recordings of a successful 'aimed' midwater tow (trawl 2,000 meshes circ) on spawning herring in the North Sea. The herring school hardly changes shape and does not show 'flight reaction' movement from trawl roping and netting. Catch, 80 tons. Note that the bottom trace in the netsonde recording indicates the vertical movement of the net. The ship's sounder echogram proves that the ground was actually flat. The midwater trawl was lowered only briefly, almost to the bottom, to scoop up the fish school and was then hauled in immediately to bring the catch on board

narrower curve and may therefore miss the school in the horizontal direction. The depth of the trawl is adjusted by varying the towing speed or length of the warps or both. The depth of the school can be estimated from the angle of tilt of the sonar and the fishing depth adjusted tentatively. This depth is confirmed by the ship's echo sounder for final adjustments. When the trawl reaches the school, last minute adjustments of trawl depth can be made by altering the towing speed. *Fig 104* illustrates a successful 'aimed' trawling attack using all three fishing aids, sonar, echo sounder and netsonde. As the catch was spawning herring, there was little change in the school shape and no indication of fish scaring or flight reaction. Another example of 'aimed' midwater trawling was given in *Fig 96*.

Pelagic fish schools vary in their reaction to vessel noise and to trawl gear. In most cases their flight is downward and the trawl depth should therefore be set to the centre or even the lower part of the school. Adjustment of towing power is then made to aim the trawl at the most dense part of the school, which may require lifting or lowering the net, depending on the evasive action which the fish take when the net hits the school. For this decisive phase the netsonde is indispensable, as it is too for towing close to the bottom (risk of damage). As an illustration, the netsonde echogram in *Fig 105* shows fish both being caught and escaping under the footrope. It also shows the trawl being manoeuvred very close to the bottom to reduce fish escapement as much as possible.

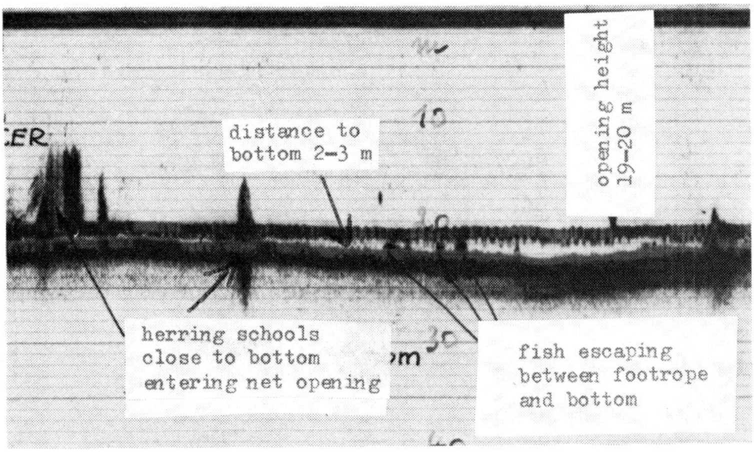

Fig 105 Netsonde echogram showing midwater trawling close to the bottom. Some of the fish still escape below the footrope

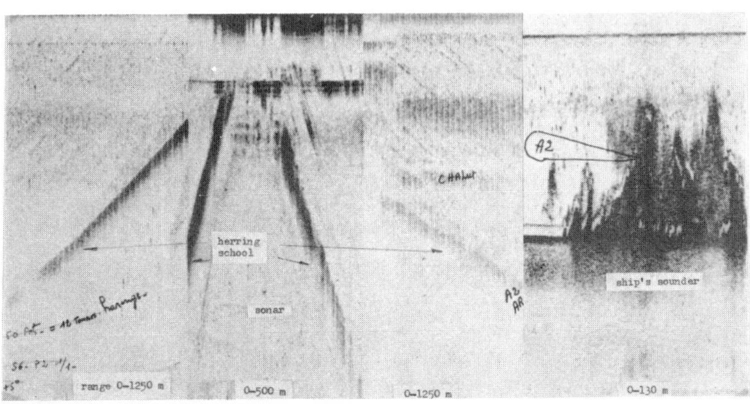

Fig 106 Sonargram made during 'aimed' midwater trawling in shallow water very close to or even touching the bottom. Under such conditions, and provided that the skipper knows the grounds and his gear very well, a netsonde is not absolutely indispensable. Note sonar observation, first ahead and then backward. Ship's sounder echogram on right, range scale 0-130m. The tow was made in the North Sea and produced 12 tons of herring

An example of operating without a netsonde is illustrated in *Fig 106*. Sonar contact was made at a range of 1 200m and the skipper changed course towards the school and shot the trawl. The distance between the vessel and school decreased evenly and therefore the echo from the school was recorded as a straight line. The range was changed and the vessel passed over the school. This was not recorded by the sonar because the transducer tilt of this set was limited to 45°. The size and depth of the school was shown on the ship's sounder echogram on the right of the figure. As the vessel passed over the school, the speed was adjusted to set the trawl on the ground and the sonar beam was directed astern. The sonar range was again increased to show the trawl and fish school on the 1 250m range. The catch from this tow was 12 tons of herring.

11.3 Purse seining

For purse seining, sonar is the main acoustic aid. The searching method and approach to the school is the same as for midwater trawling, except that the vessel usually does not pass over the school (particularly when it is close to the surface) to avoid undue excitation of the fish. The vessel is brought to a range from the school which is approximately the radius of the purse seine. The net is then shot keeping the vessel's heading at about 90° from the bearing of the school and at a constant range. *Fig 107* is a typical example of a sonargram obtained from this technique.

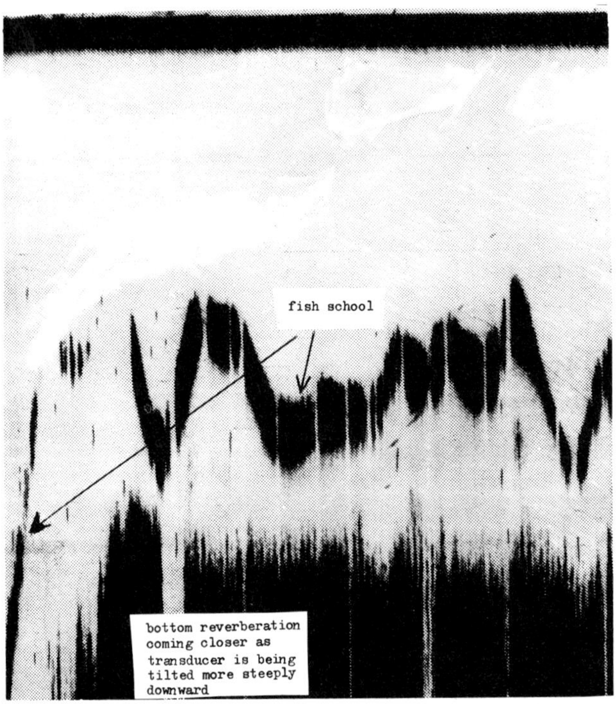

Fig 107 Sonargram of a fish school maintained at approximately constant distance from the vessel, which is circling round it for assessment and then setting the purse seine

If the school moves significantly and if there is some current, shooting of the net normally starts in the direction the school appears to be moving or towards the current (*Fig 108*), otherwise the fish would escape or the current might take the net over the school before it has time to sink sufficiently. The sonargram in *Fig 109* refers to that of *Fig 103* (searching) showing the tuna circling in the closed purse seine.

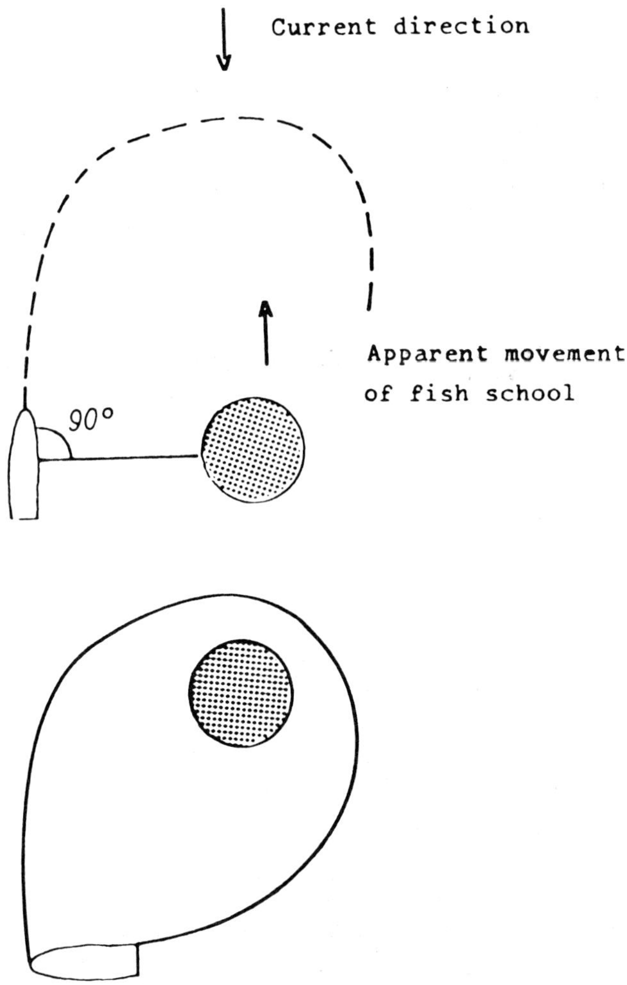

Fig 108 A typical sonar-guided purse seine set, made with the current and against the apparent movement of the fish school. Top: Relative position of vessel and school when shooting starts. Below: Taking the buoy. The current has carried the centre of the net towards the school

11.4 Rock and wreck fishing

This method of fishing may involve gillnets, longlines, handlines or pots being set close to or across a known wreck or rock (*Fig 110*) and it is based on the observation that fish tend to congregate around such objects on the

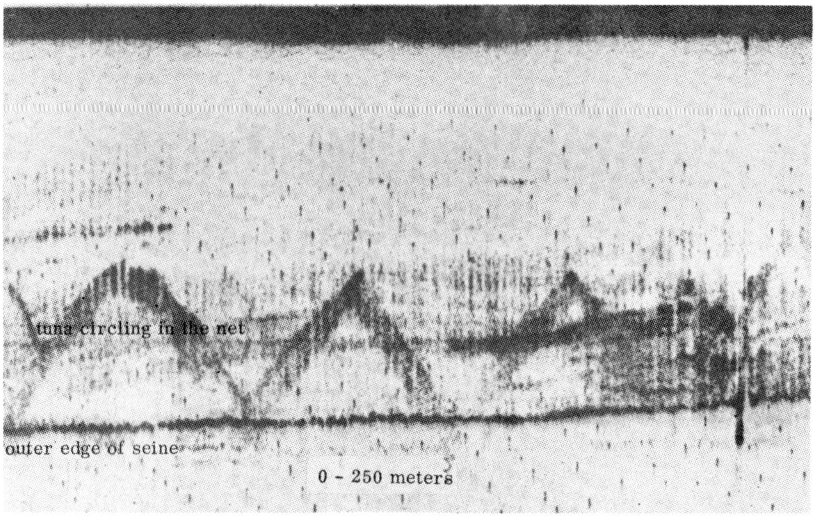

Fig 109 Sonargram showing tuna circling in a closed purse seine

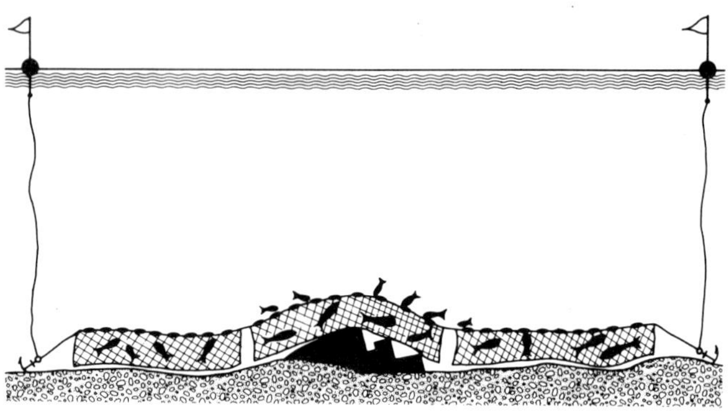

Fig 110 A practical example of setting gillnets across a wreck

sea bed. The wreck or rock position is usually marked on fishing charts or known to the skipper and a suitable navigational aid is used to reach the correct area. The sonar is then set to search over a sector of about 120° in the forward direction, with a fixed tilt angle and a short range setting. When the object is picked up the vessel is headed straight for it. The skipper may first run over it to ascertain his position accurately with the ship's echo sounder and to assess its actual fishing potential. He then decides on exactly where and how to place his gear. As an example *Fig 110* illustrates a set of three gillnets, 50m each, joined together and equipped with anchors and dhan buoys. These nets are usually set across the tide so that the middle net is across the rock or wreck. Two or three such sets of nets may be laid, depending on the size and type of the obstacle.

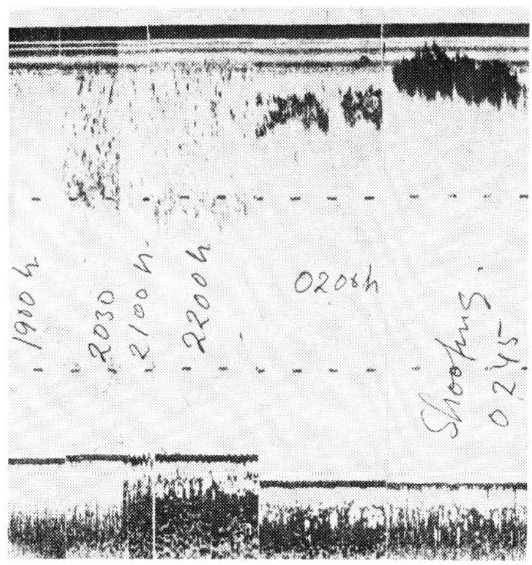

Fig 111 (Right) Echogram showing slow, progressive, light attraction of sardinella by a purse seiner. After almost eight hours, only a small amount had concentrated, yielding a catch of about 500kg

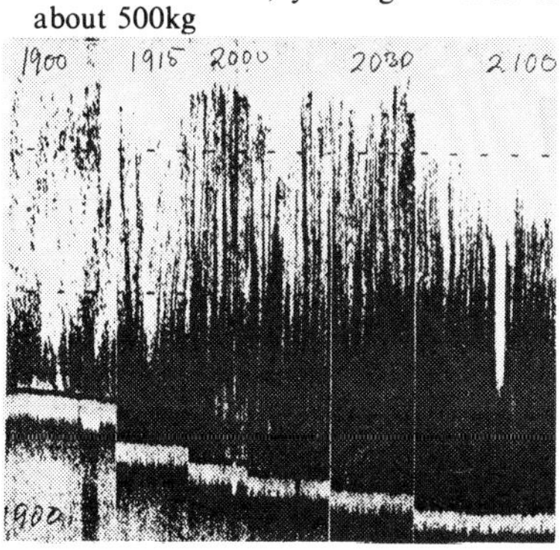

Fig 112 (Above, left) Echogram showing very successful light attraction of sardinella by the same vessel. This was within one week (mid-May 1979) of the example given in *Fig 111* and in the same general area of the northern part of the Gulf between Iran and the Arabian Peninsula. The changes in water depth in between checks show that the vessel was drifting. After only two hours of light attraction, the catch was about 35 tons

11.5 Fish attraction

Two common methods of attracting and concentrating fish are, artificial light and floating lures (rafts). The main techniques employed for catching fish, or other organisms such as squid, which have been attracted in this way are purse seining or lampara seining, lift netting, hand lining or jigging and even pumping. A great variety of pelagic fish ranging from small bait fish (attraction by light) to tuna (lures, rafts) can be concentrated in this way. Echo sounding, and to some extent sonar, serve to find suitable areas where target fish are dispersed and to observe the concentration process as well as to judge quantity and position in order to decide whether to fish and when.

As an example of purse seining for sardinella with light attraction in the Gulf between Iran and the Arabian Peninsula *Figs 111* and *112* show echograms from which the progress of fish concentration could be observed. The position and amount of fish is shown clearly and is in obvious accordance with the catches obtained, *ie* 500kg in the first case and 35 000kg in the second.

11.6 Bottom trawling with sonar

The combination of echo sounder and sonar can be employed to great advantage to assess and exploit difficult and rough (un-trawlable) grounds using the echoes from the sea bed on the sonar to control the path of the

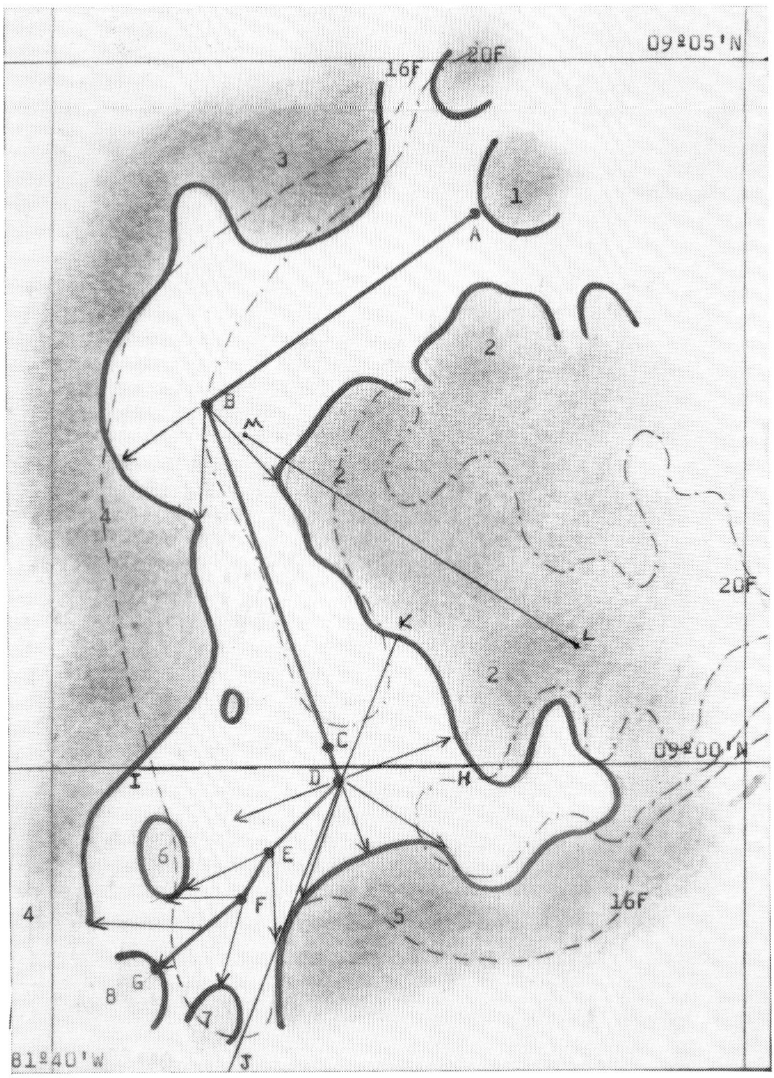

Fig 113 Shrimp trawling in a coralline area of the Western Caribbean Sea by an FAO Project vessel, using sonar. This difficult ground was trawled for 90 hours, without gear damage. A regular night-time catch by the two 54ft shrimp trawls was 500-600lb of pink shrimp *(P duorarum)*, heads-off, whereas the nearby clear trawling grounds were producing 300-400lb. The former ground was not being exploited commercially

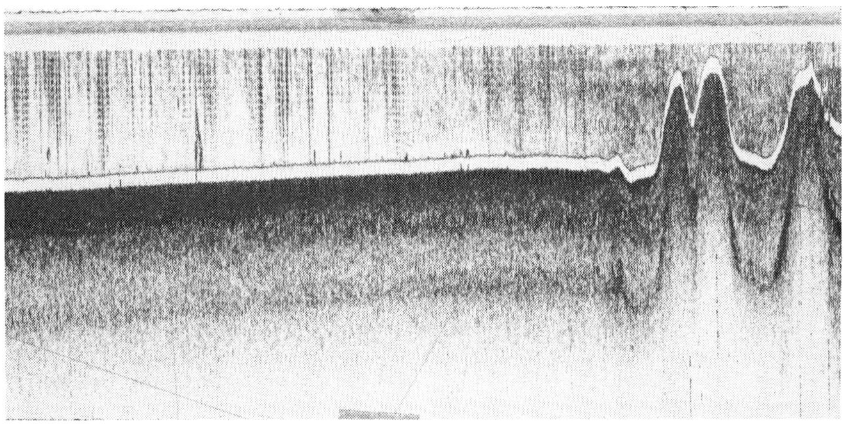

Fig 114 Ship's sounder echogram of seabed between points 'H' and 'I'

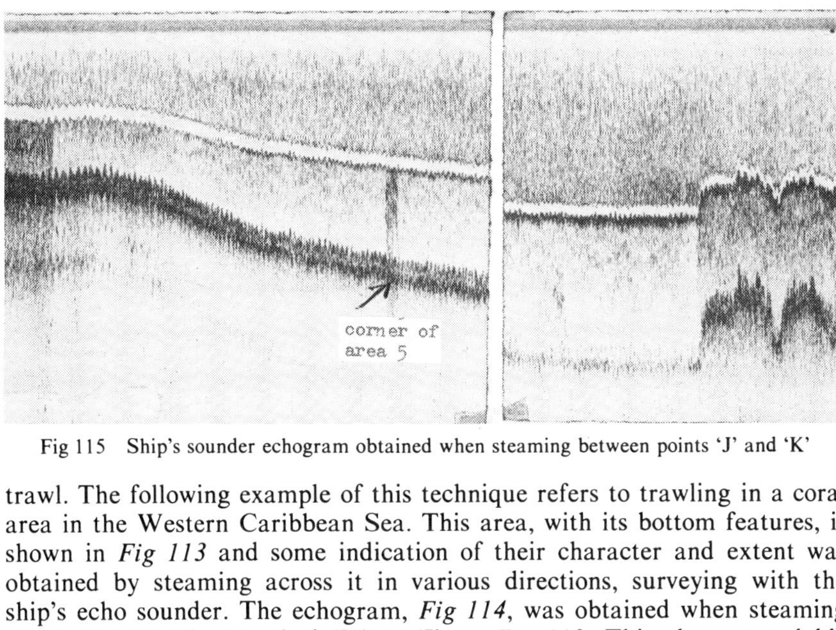

Fig 115 Ship's sounder echogram obtained when steaming between points 'J' and 'K'

trawl. The following example of this technique refers to trawling in a coral area in the Western Caribbean Sea. This area, with its bottom features, is shown in *Fig 113* and some indication of their character and extent was obtained by steaming across it in various directions, surveying with the ship's echo sounder. The echogram, *Fig 114*, was obtained when steaming along the direction marked 'H' to 'I' on *Fig 113*. This shows trawlable ground, which is regular, smooth and sandy and at a depth of 17fm, until reaching the steep ascent to the rough area 4 with a depth of only 6fm of water over it. The fact that there are no strong traces of second and third bottom echoes confirms the area 4 to be coral rather than hard rock.

Fig 115 is the echogram obtained when steaming along the line 'J' to 'K'. This path starts from the rough area 7, with flat corals at a depth of 9fm, and ends at rough area 2, also with coral, at an approximate depth of 15fm.

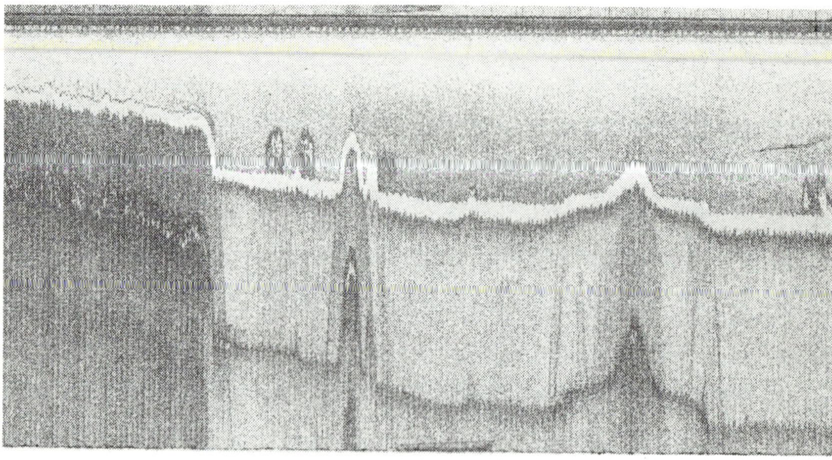

Fig 116 Ship's sounder echogram obtained between points 'L' and 'M'

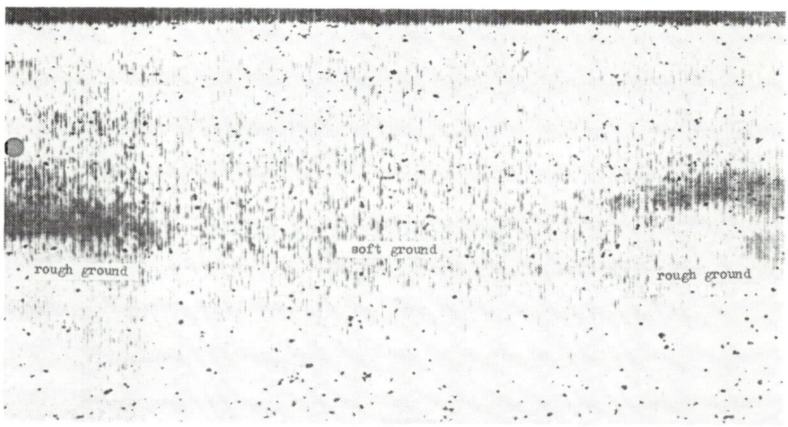

Fig 117 Sonargram showing distance and bearing to the edge of rough, non-trawlable ground

Both these areas are clearly indicated by the increased bottom echo length at both ends of the echogram. While travelling over the trawlable ground, the vessel passed close to rocks (5) as indicated by the increased echo length produced by the transducer side lobes.

A check on the sea bed in the centre of the area is shown on the echogram, *Fig 116,* taken when steaming between 'L' and 'M'. At the start the bottom is hard, compact, coral and rocky, followed by a mixed area of sand with coral outcrops; these latter are indicated by oblique traces from the side lobes and partly even by whitelining of the second bottom echo. Finally there is a coral spot before dropping to the sandy trawlable area on the right hand side of the echogram.

To exploit the ground, trawling commenced at point 'A' along the path

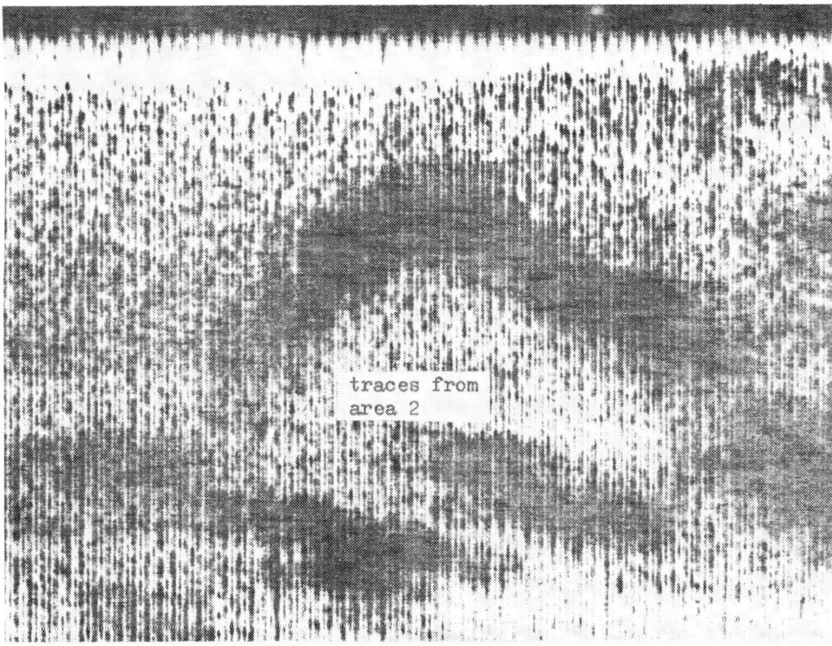

Fig 118 Sonargram showing rough bottom, close to one side of the trawler towing along the edge of the area marked '2' *(Fig 113)*

'AB' between rough areas 2 and 3, which were shown on the sonar to be separated by approximately 1 mile. While trawling, the sonar beam was directed to the port side in order to maintain the vessel's position relative to the edge of the rough area 2. The distance was held at approximately ¼ mile and the edge of the rough ground is clearly shown by the black bottom traces on the sonargram (*Fig 117*). The beam was periodically trained ahead and to starboard in order to check the ground ahead and the distance clear from rough area 3. This was necessary to avoid any unexpected rocky outcrops.

After some time of towing the sonar detected the rough area 4 dead ahead and a free channel between areas 2 and 4 in a southerly direction. The course of the vessel was altered at point 'B' and set to follow the line 'BC' keeping close to area 2. At the beginning (*Fig 118*), the edge of the rocks was first detected at a distance of ⅓ mile. The beam was maintained on this edge and the black approaching bottom trace on the right hand side shows a nearest point of approach of ⅕ mile.

Having safely passed through the rather narrow gap between rough areas 2 and 4, the sonar beam was predominantly directed ahead, and a rather strong echo was picked up at maximum range (*Fig 119*). The fact that this was obtained from a large distance would suggest a possibly rocky obstruction. This trace was therefore closely observed but from time to time the beam was transferred to check the distance from area 2, which was held at approximately ⅓ mile (first two dark spots on the sonargram). Similarly a check was made to starboard on area 4 (third spot on the sonargram). As the vessel approached the apparent obstruction dead ahead, it can be seen that the echo strength decreased and the trace almost disappeared.

The target therefore, was clearly not rock on the sea bed. It was, in fact,

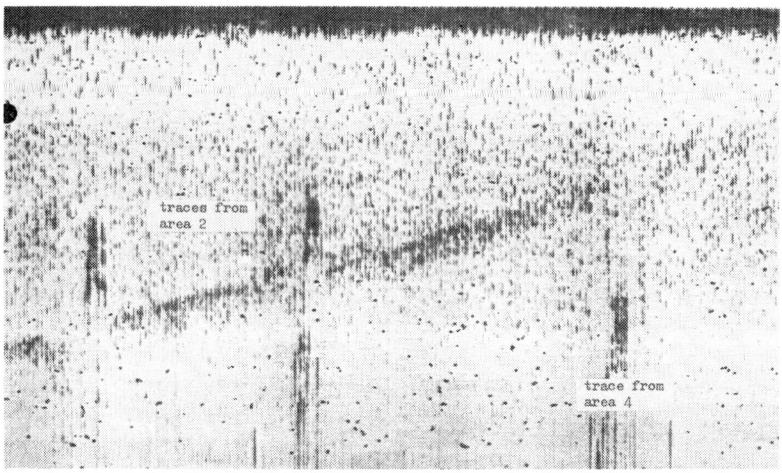

Fig 119 Sonargram obtained when trawling along the line '**B-C**'

found to be a water discontinuity on the surface; blackish, muddy water was invading the normally clear main body of sea water. A similar effect may be obtained from drifting debris or seaweed. Such targets are identified by the fact that the echo strength reduces and the trace even disappears as the object is approached, that no shadow sector appears behind the trace and that the audio signal remains low and soft.

Following this path along 'BC', it was first observed that the channel had widened, since rock echoes were no longer obtained from either side of the trawler's course. Finally the true obstruction of rough area 5 was detected. As there was sufficient manoeuvring room around the vessel, the course was not immediately altered and a better observation of area 5 was obtained. There were two possible free channels to select; one between rough areas 2 and 5 and the other between rough areas 4 and 5. The latter was chosen and the course altered at 'D'. At point 'E', the rough spot 6 was detected with a possible free passage between 6 and 5. The course was altered at 'E' to take the middle of this channel until point 'F' when rough spot 7 was detected on the sonar. The free channel between areas 6 and 7 was taken until rough spot 8 was detected dead ahead. This appeared to be the southern limit of the trawlable grounds so the trawl was hauled in. The end phase of this 7 mile tow is shown on the echogram *Fig 120*.

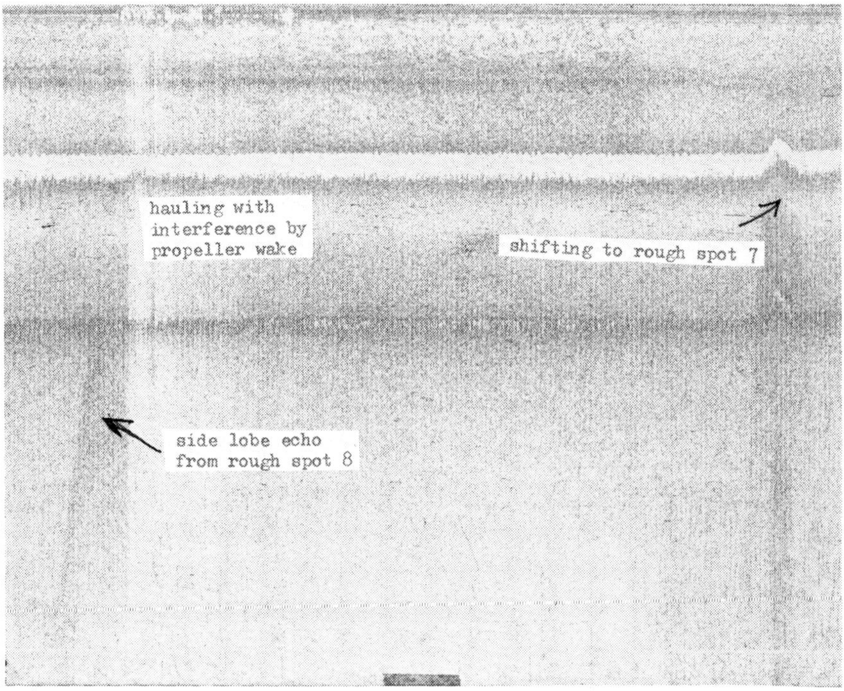

Fig 120 Hauling at the end of the tow when the rough spot at '8' appeared, with side lobe traces on the echogram of the ship's sounder

Trawling continued straight towards the rough spot 8 until it appeared on the ship's sounder as oblique side lobe traces. The propeller was then put into neutral and the trawl hauled. The trawl acts as a sea-anchor in these circumstances and the vessel was pulled astern away from the rough spot. In this way, the maximum trawling distance was obtained without danger to the gear. As the vessel goes astern, the acoustic beam is disturbed by the wake so giving changes in the noise and sea bed echoes as well as the zero line traces.

After hauling and while preparing for the next shot, the vessel headed slowly towards rough spot 7 and, when this appeared on the echogram the gear was shot in the opposite direction, again maximizing the trawling distance.

CHAPTER 12

FUTURE TRENDS

It will be apparent from previous chapters, that improvements to existing equipment are required; for example the need for a narrow beam transducer, for a higher rate of obtaining target information on sonars and a reduction in the effort required of a skipper in interpreting the displays. Manufacturers and scientists are working on these problems, and improved equipment is becoming available. Many improved installations at similar or even reduced price have been made possible by the rapid development of semi-conductors and integrated circuits.

Apart from the more sophisticated acoustics mentioned below, efforts are also being made to benefit from this technical progress by developing low cost echo sounding equipment suitable for medium and small scale fisheries in developing countries. The performance of this future cheap, rugged and reliable equipment will be more or less equal to the small to medium size units discussed earlier. Even if modified display systems should come into use the contents of this manual will remain relevant for assisting newcomers in using the technique to best advantage.

12.1 Non-linear acoustics

The value of a narrow acoustic beam has been pointed out (*4.2*) and if in addition the beam had no side lobes, then the bottom reverberation experienced with sonar (*8.2*) would be reduced. Using standard transducer and acoustic techniques, a narrow beam can be obtained only from a large transducer, with shading or tapering to reduce the side lobes. The size of this transducer can be reduced to a more manageable size by operating at a higher frequency, but this in turn reduces the range capability of the set, since power transmitted at higher frequencies is attenuated more rapidly. By exploiting the fact that the propagation of acoustic waves in the sea is non-linear, it is possible to transmit two frequencies coherently in such a way that they interact to produce a wave at a frequency which is the difference of the two primary frequencies. Thus a specially designed multi-element transducer array can have alternate elements simultaneously energized by two high frequencies and therefore be small in size but can project a narrow beam, low frequency wave in the water. The interaction can be regarded as forming a long 'end-fire array' which is highly directional and without side lobes.

12.2 Within Pulse Electronic Sector Scanning (WPESS)

The slowness of the mechanical scanning of the sonar transducer prevents the use of narrow beams with which high angular resolution could be obtained. As mentioned under *9.4*, omni-directional transmission with narrow pre-formed beams for reception and electronic scanning and storage assist with this problem, but it can be overcome by a sophisticated system known as Within Pulse Electronic Sector Scanning Sonar (WPESS). In this system, a wide sector is insonified (irradiated by sound) and a narrow receiving beam is swung electronically so that the whole sector is examined for echoes before the transmitted pulse has travelled its own length in the water. An equipment in use for fisheries research operates at 300kHz, transmits over a 30° sector, which is then scanned at a rate of 10 000 times/second using a transducer with a receiver beam width of 0.33°. Such a narrow beam has to be operated from a stabilized platform and *Fig 69* illustrates the use of this type of sonar in fishing gear research.

This sonar is too expensive for commercial fishing, but it is expected that, with the use of micro-electronic techniques and a simplified approach, a version will be produced at an acceptable price. Already one set can survey 360° at a rate of 33 000 rev/min giving a new display of the area to a range of 1,900m every 2½ seconds. A PPI display is used with a special spiral scan of the tube.

12.3 Television type displays

The main advantage of the CRT type of display is that there are no mechanical moving parts, no need to change paper or stylus and therefore a more reliable set can be produced. The disadvantage is that there is no record to which the skipper can refer when determining the tactics for the next or future operations. As micro electronics are improved, it has become easier and more economic to store and process information. For a reasonably low price, a square tube echo sounder is already available which provides a black and white CRT display equivalent to a paper record of 135mm length. A much more expensive colour TV echo sounder display with electronic memory and signal processing has also recently become available. A Polaroid camera is provided for instant picture recording and a cassette tape recorder allows replay of any required sequence. The presentation gives strong echoes of the bottom in red, shifting colour through to blue for echoes of less intensity such as fish.

12.4 Multiple information on a single display

As discussed in Chapter 11, for midwater trawling a combination of sonar, echo sounder and netsonde is used to detect and follow the school of fish and to position the trawl. The displays of these sets could in future be combined into one. For example a dual frequency echo sounder is available which has a chart display with the normal white line, scale expansion and bottom or transmission lock facilities, but on which the information from a netsonde is also displayed. This latter information is telemetered from the trawl and consists of echoes from above and below the trawl headline.

Another approach to presenting information for midwater trawling or purse seining operations is Situation Display. This system consists of a mini-computer to which are connected a sonar, a speed log and a compass, and the information is displayed on a CRT showing the position of the fish school, its size, depth, speed and direction of movement, as well as the track of the ship, its speed and heading and the distance to the school. Such a system is particularly useful in purse seining, where the operator sets a small square symbol on the screen to indicate the position of the end buoy and the computer controls the sonar transducer to track a pre-selected school of fish and the relative movements of the vessel.

12.5 Counters

Echo sounders are also used by scientists for the acoustic assessment of fish stocks or biomass. This has given rise to a variety of fish counting and echo integrating techniques with scientific echo sounders equipped with correct TVG.

A towed body may be used to house the transducer and so reduce vessel interference and obtain fish echoes up to the surface. Numbers of fish and their target strength for various depth bands may be processed by computer. Echo integration, adding together all the echoes returning from a pre-set depth band over a fixed time or distance travelled, enables us to calculate the fish quantity per volume of water if the target strength of the fish recorded is known. One approach investigated for the application of this technique to commercial bottom fishing has been to count the returns from fish in a pre-set depth band, then correct this count for depth variations and multiply by a factor established from the results of the previous tow, in order to estimate the probable catch for the present tow.

One of these counter or integrator systems may be developed further so that the difficulty of estimating the probable catch from echo sounder or netsonde records can be overcome.

FURTHER READING

As pointed out in the Introduction, this manual concentrates on practical applications in fishing. For those who wish to study underwater acoustics and its application, not only to fishing, but also to other related research and development purposes (*eg* studies on fishing gear and methods, fish behaviour, fish migrations, resources assessment) more thoroughly, the following list of bibliographic references provides a choice of books and papers, most of which make reference to further, often more specialized, publications. Quite a few relevant papers were submitted to the three FAO organized Fishing Gear Conferences and, together with discussions, are contained in the respective volumes of Modern Fishing Gear of the World (Vol I, 1959; Vol II, 1964; and in particular Vol III, 1971, all co-published for FAO by Fishing News Books Ltd).

AZHAZHA, V G and SHISKOVA, E V *Fish location by hydroacoustic devices.* Jerusalem, Israel 1967 Programs for Scientific Translations, IPST Cat No 1856: 114pp, 92 figs, 20 tables

BURCZYNSKI, J, *Introduction to the use of sonar systems for estimating fish biomass.* FAO Fish, Tech. Pap., (191): 89p

CUSHING, D H, *The detection of fish.* Oxford, Pergamon Press Ltd, 200pp, 94 figs 1973

DUNLOP, G D, *Navigation and finding fish with electronics.* Camden, Maine, International 1972 Marine Publishing Co, 20pp, figs, 5 tables

ELAC *An introduction to echosounding.* Kiel, Electroacustik GmbH, 146pp, 35 figs. 1965

FORBES, S T and NAKKEN, O (Ed), *Manual of methods for fisheries resource survey and* 1972 *appraisal, Part 2. The use of acoustic instruments for fish detection and abundance estimation.* Rome, FAO, *FAO Man Fis Sci*, (5): 138pp, figs

HAINES, R G, *Echo fishing.* London, Leonard Hill, 116pp, 15 figs, 54 plates 1969

HAINES, R G, *Sound under water.* Newton Abbot, David and Charles, 208pp, 17 figs, 12 plates 1974

KRISTJONSSON, H, *Techniques of finding and catching shrimp in commercial fishing.* Rome, 1968 FAO, *FAO Fish Rep*, (57) Vol 2: 125–192, 57 figs, 1 table

SCHÄRFE, J, *The German one-boat mid-water trawl.* London, Fishing News International, Vol 1969 8, No 7 26–33, No 8 18–22, No 9 34–41, No 10 20–23, No 11 27–35, No 12 36–43; also as reprint, 29pp, 28 figs

SIMRAD, *Fish-finding with sonar.* Oslo, Simonsen Radio AS, 96pp, figs 1965

SIMRAD, *Pop technique. The echo sounder.* Oslo, Simonsen Radio AS, 28pp, figs 1971

TUCKER, D G, *Underwater observation using sonar.* London, Fishing News Books Ltd, 144pp, 1966 51 figs

TUCKER, D G, *Sonar in fisheries—a forward look.* London, Fishing News Books Ltd, 136pp, 1967 57 figs

TUCKER, D G and GAZEY, B K, *Applied underwater acoustics.* Oxford, Pergamon Press Ltd, 1966 244pp

URICK, R J, *Principles of underwater sound.* New York, McGraw-Hill Book Company, 384pp, 1975 figs and tables

WOOD, W J, *The echo sounder and its application.* Grimsby, British United Trawlers Ltd, 1972 21pp, 12 figs

YUDANOV, K I, *Interpretation of echograms of hydroacoustic fishfinding instruments.* 1971 Jerusalem, Israel Program of Scientific Translations, 101pp, 60 figs

Fisheries periodicals inform about new developments by articles and also by the advertisements of manufacturers, suppliers and agents. Another obvious source is technical leaflets and instruction manuals, which suppliers will provide on request or which can be collected at fisheries exhibitions. Some of the larger manufacturers even have their own house journal to which one can subscribe. New findings and developments are often submitted as papers to such meetings or working groups as are organized by a variety of fisheries bodies. These may not be easily accessible but could eventually be obtained through the respective national fisheries services, particularly if these include an officer or a group specifically concerned with fishing technology.

Obviously, books and papers are essential to provide background knowledge, guidance and stimulation, but these need to be complemented by practical experience, which can only be obtained through attentive 'learning by doing'. Operational efficiency depends on both theoretical knowledge and its rational application. As in all technology the best equipment cannot perform better than the man who operates it.

Other books published by Fishing News Books Ltd Farnham, Surrey, England.

Free catalogue available on request